The Plausible World

ALSO BY BERTRAND WESTPHAL

Geocriticism: Real and Fictional Spaces

La Géocritique. Réel, fiction, espace

Austro-fictions. Une géographie de l'intime

L'œil de la Méditerranée. Une odyssée littéraire Roman et Évangile

Littérature et espaces (with Juliette Vion-Drury and Jean-Marie Grassin)

La géocritique mode d'emploi (editor)

The Plausible World
A Geocritical Approach to Space, Place, and Maps

Bertrand Westphal

Translated by Amy D. Wells

THE PLAUSIBLE WORLD
Copyright © Bertrand Westphal 2013.

All rights reserved.

First published in 2013 by PALGRAVE MACMILLAN® in the United States—a division of St. Martin's Press LLC, 175 Fifth Avenue, New York, NY 10010.

Where this book is distributed in the UK, Europe and the rest of the world, this is by Palgrave Macmillan, a division of Macmillan Publishers Limited, registered in England, company number 785998, of Houndmills, Basingstoke, Hampshire RG21 6XS.

Palgrave Macmillan is the global academic imprint of the above companies and has companies and representatives throughout the world.

Palgrave® and Macmillan® are registered trademarks in the United States, the United Kingdom, Europe and other countries.

ISBN: 978-1-137-36458-6

Library of Congress Cataloging-in-Publication Data

Westphal, Bertrand.
 [Monde plausible. English]
 The plausible world : a geocritical approach to space, place, and maps / Bertrand Westphal ; translated by Amy D. Wells.
     pages cm
 Includes bibliographical references and index.
     ISBN 978-1-137-36458-6 (hardback : alk. paper) 1. Geography—Philosophy. 2. Geography in literature. I. Wells, Amy D., 1974- II. Title.

G70.W4713 2013
910'.01—dc23   2013024513

A catalogue record of the book is available from the British Library.

Design by Scribe Inc.

First edition: November 2013

10 9 8 7 6 5 4 3 2 1

To Liv,

who runs the whole world over with great strides

Contents

Translator's Note	ix
Foreword: A Geocriticism of the Worldly World *Robert T. Tally Jr.*	xi
Acknowledgments	xvii
Introduction	1
1 The Multiplication of Centers	9
2 The Horizon Line	43
3 The Spatial Urge	73
4 The Invention of Place	103
5 The Measured Mastery of the World	135
Notes	165
Index	185

Translator's Note

To my fellow geocritical readers,

Prepare yourself for the journey that is this text. We will be traveling across four thousand years (at least) of history and to all corners, and more important, to the multiple centers of the globe. Professor Bertrand Westphal executes his geocriticism coming from the French discipline of Comparative Literature, and, on top of that, he is a passionate world traveler. It is for these reasons that works from a dozen or so literary traditions are cited and maps from a multitude of cultures are evoked. Your excursion will take you not only to Paris, France, but also to Paris, Texas.

Given the rich cultural and literary diversity of this study, important translation choices have been made. Where possible, the original English versions of sources are cited. Also when possible, previously published English translations of canonized texts are used. If no prior translation of quoted secondary matter exists or was available, I have myself translated it for this monograph.

Professor Westphal's writing style, like that of many French critics, reposes on wordplay. Some expressions are untranslatable, and when the wordplay contributes significantly to the meaning of the text, there is a translator's note to explain the nuanced meanings. There are also occasional translator's notes to guide readers regarding specific cultural references throughout the text when a word used is indicative of a social class, region, and so on. Furthermore, numerous place names are employed, and they are our landmarks as we move through space to place using maps. Bear in mind that these place names are of historical, political, and mythological origins: some may have contemporary Anglicized versions, whiles others may not. And let us not forget that geography, cartography, and navigation are sciences with specific technical terms. Equivalents of these have been found in the majority of cases, but they may seem unfamiliar, as they are not a part of our everyday English.

As he retraces the history and the lines of space, place, mapping, literature, and literary mapping in image and in text, Professor Westphal pays particular attention to the etymology of words, either in the body of the text or in his footnotes. I have tried to pay equal attention to his analysis in providing translations that help readers understand the ways in which words have evolved and have found their place in the geocritical lexicon.

Bearing these linguistic specificities in mind, I now invite you to enjoy your trip to the Plausible World.

A. Wells
Cherbourg, France

Foreword

A Geocriticism of the Worldly World

Robert T. Tally Jr.

After what has been termed the "spatial turn" in literary and cultural studies, critics have increasingly focused attention on the relationships among space, place, and mapping in literature. This has resulted in a growing body of critical scholarship in the interdisciplinary field of spatiality studies, broadly defined so as to encompass geocriticism, geopoetics, and the spatial humanities. Recent examples of work in this field would include my *Spatiality* (2013) and *Geocritical Explorations: Space, Place, and Mapping in Literary and Cultural Studies* (2011); Eric Prieto's *Literature, Geography, and the Postmodern Poetics of Place* (2013); Amy Elias and Christian Moraru's edited *The Planetary Turn: Art, Rationality, and Geoaesthetics in the Twenty-First Century* (2013); Peta Mitchell's *Cartographic Strategies of Postmodernity: The Figure of the Map in Contemporary Theory and Fiction* (2010); David J. Bodenhamer, John Corrigan, and Trevor M. Harris's *The Spatial Humanities: GIS and the Future of Humanities Scholarship* (2010); and the massive collection edited by Michael Dear, Jim Ketchum, Sarah Luria, and Doug Richardson, *GeoHumanities: Art, History, Text at the Edge of Place* (2011), to name only a few notable books. That this body of research appears to be growing at an almost exponential rate is evidence of the timeliness of spatiality. As director of the Espaces Humains et Interactions Culturelles research group at the University of Limoges, Bertrand Westphal has been among the leading figures in this burgeoning area of inquiry.

In his 2007 study, *La Géocritique: Réel, fiction, espace* (translated into English as *Geocriticism* in 2011),[1] Westphal outlined and argued for a geocentric approach to general and comparative literature. A literary or cultural critic using this approach would begin by focusing attention on a singular geographical place, such as a city or a body of water, as opposed to a particular author, literary genre, or historical period. The geocritic would then examine the ways in which that place has been represented in a variety of texts, which could include not only works of literature proper but also films, travel narratives, governmental surveys, tourist brochures, and so forth. For example, Westphal has made a geocritical study of the Dalmatian islands, examining the literary representation of that geographical domain.[2] The geocritical method advocated by Westphal would allow readers to see the ways in which various texts represent the spaces of the locale selected, and by examining a variety of perspectives, geocritical scholars could develop a relatively nonbiased, though inevitably incomplete, image of the place. One serious problem for those wishing to employ this geocritical method, a problem readily acknowledged by Westphal, is that it invariably raises the question of the corpus. How does one determine exactly which texts could, in the aggregate, reasonably constitute a meaningful body of material with which to analyze the literary representations of a given geographical site? That is, if the Dublin of James Joyce is far too limited, since it relies on the perspective of a only single author or a few of his own writings, then how many authors and texts representing Dublin would constitute a feasible and credible starting point for a geocritical study of the Irish capital? With certain cities, such as Paris, London, Rome, or New York, the almost mythic status of these places and the seemingly innumerable textual references to them render any geocritical analysis, at least those laying claim to a kind of scientific value, impossible. As Westphal admits, "to attempt to undertake a full-scale geocritical analysis of those hotspots would be madness."[3] A geo-centered method, if it aims truly to avoid the perception of bias, seems somewhat doomed from the start.

Another potential problem with Westphal's geocentric method, one that Prieto has addressed, is that it does not adequately take into account the conflicting forces and views that condition the ways in which various spaces come to be recognizable as places. Prieto notes that Westphal's focus, perhaps understandably, has been on the "*hauts lieux* of the literary tradition: places that have a distinct cultural and topographical profile and that have given rise to a whole body of literature."[4] In his revision of the geocritical project, Prieto considers the textual emergence of various types of place, including what appear to be nonplaces, such as improvised shantytowns or the French *banlieues*, along with already identified, recognizable *topoi*. By examining the ways that certain kinds of social space come into being as places for interpretation, this sort of geocriticism offers intriguing opportunities for understanding how literary representation and spatiality interconnect. Or, to state it somewhat differently, geocriticism would need to examine not only the real and imagined places of literature but the conditions for their possibility as well.

In *Le Monde plausible: Espace, lieu, carte* (2011), which appears here in English thanks to Amy D. Wells's translation, Westphal demonstrates the degree to which he too was puzzling through such problems. As a follow-up to his more programmatic and introductory *Geocriticism*, the present volume explores a rich variety of texts, from antiquity to the present, drawing on numerous cultural and linguistic traditions from all over the planet. In this extravagant, sometimes sprawling, but also intensely curious performance, Westphal explores a wealth of material in attempting to characterize the many ways in which individuals and cultures enact a sort of literary cartography of their worlds. Westphal argues that a certain kind of cartographic practice, associated with a rational or scientific discourse, developed in the West. It is embodied perhaps most vividly by the networks of imaginary lines—latitudinal and longitudinal—that imprison the spaces of the world in a Cartesian grid. However, at the same time alternative mapping practices arose and flourished in other, non-Western civilizations, and Westphal envisions these countermaps as implicit, and occasionally explicit, challenges to the hegemony of the more limited Western spatial imagination.

In my own opinion, I believe that Westphal's use of the unproblematized geopolitical metaphors of the West and non-West is itself rather problematic and sometimes smacks of a discourse of Orientalist oversimplification, as when Michel Foucault notoriously contrasted a "Western" *scientia sexualis* with the "Eastern" *ars erotica*.[5] I think that it should be obvious in an age of globalization, if not much earlier, that the skein of power-knowledge relations that makes possible the imaginative cartography of our world is far more complicated. As I think the historical record bears out, even the most hegemonic mapping practices, like most practices, in fact, entail apparently counterhegemonic ones, and vice versa. For instance, the imposition of a grid-like plan for an urban space has never really stopped the wayward *flâneur* from taking a shortcut, perhaps even a non-Euclidean shortcut; but it is equally clear that errant wanderers who make their own paths are frequently discovered to be pathfinders for an even more rigid stratification, as today's shortcut becomes tomorrow's one-way street. The geocritic does well to remember Gilles Deleuze and Félix Guattari's discussion of smooth and striated space, a distinction that partly underlies Westphal's own views: "What interests us in operations of striation and smoothing are precisely the passages or combinations: how the forces at work within space continually striate it, and how in the course of its striation it develops other forces and emits new smooth spaces." As Deleuze and Guattari note, "even the most striated city gives rise to smooth spaces."[6]

Notwithstanding my objection to what I consider an oversimplified and perhaps overly moralizing conception of the West and its "others," the overall project of *The Plausible World* is entirely worthwhile. Westphal's comparative literary treatment of spatial representation is both fascinating and stimulating, with interpretations that will encourage readers to read more and to read otherwise. By moving away from the strictly geocentric approach outlined in the methodological sections of *Geocriticism*, Westphal connects his critical practice to a broader spatiality studies that does not resist the allure of an individual author's perspective, while also paying closer attention than in the former study to the problem of the *emergence* of places. Hence *The Plausible World* nicely extends the project of *Geocriticism* and, at the same time, strikes out in interesting new directions in search of unexplored territories.

In this book, Westphal challenges the view that perceptions and representations of space are largely stable and straightforward. For example, although maps often give the impression that the geographical knowledge of the world at any given moment is complete, that the planet has lost any sense of wonder now that all that was to be discovered has been not only found but mapped, Westphal believes that this is misleading and that it is characteristic of a peculiarly Western modernity.[7] Throughout its history, in Westphal's view, the West—through direct imperialism and through scientific practices—has repeatedly confronted open spaces of the world and transformed them into closed places. But enclosures of such places have never been definitive or unchanging. Westphal finds in the non-Western arts a countergeography to the cartographic pressures of the Western modernity. The visual art of the Aztecs, the cartographies of the Far East, and the chanted lines of Australian Aborigines confirm that the West never held a clear monopoly on geographic images of the world. These twists and turns, through spaces and places of the past and present, postulate the existence of, not just the one "real" world or infinite possible worlds, but a *plausible world*, a new conception of *the* world as *plausible*. Acknowledgement of the merely plausible world, as Westphal sees it, would spell the end of the hegemonic claims of the West. The geocritical exploration of these alternative ways in which spaces and places have been imagined not only leads to new insights into our world but discloses hitherto invisible or unknown elements that a purely Eurocentric model has left hidden or ignored. In a sense, Westphal's book aims to make visible a new world.

Drawing on his background in classical literature, Westphal begins *The Plausible World* with an examination of ancient Greek mythology, particularly looking at the conception of an *omphalos*, "navel," or center of the world. From this point of departure, Westphal gradually proceeds through various traditions of spatial representation up through modernity and across multiple cultural formations. In Westphal's view, the modern West attempted to master space, and the exploration and colonization of new spaces by Columbus and those who followed formed part of this cartographic program. However, as Westphal also insists, beginning with Cabeza de Vaca and continuing through twentieth-century "Third World" writers and artists, alternative perceptions and depictions of space emerged.

Westphal attempts to eschew a Eurocentric perspective by looking at various non-Western traditions, such as those to be found in Africa, Australia, and China. He argues that the ethnographic (and ethnocentric) character of Western spatial sciences must be replaced by a multicultural and multifocal perspective. Westphal draws heavily on postcolonial theory in his geocritical reading of the texts under consideration, and he introduces such lesser known practices as Aztec mapmaking, transoceanic voyages by explorers from China and Mali, or aboriginal representations of space. These "alternative spatialities" allow geographically oriented critics to rethink the ways they imagine the world. The alternatives do not, and ought not to, simply replace the Western models, nor do they somehow reveal a "true" world that the ideologically suspect West presented falsely. Rather, the critic's engagement with multiple spatialities makes possible a *plausible* world that does not claim for itself immutable, apodictic reality. Westphal's commitment to the plausible world, as opposed to the more definitive singular (*the* world) or to loose pluralities (of possible or multiple worlds), indicates the flexible and still tentative view of such a far-reaching and theoretically ambitious geocritical project. Along the way, *The Plausible World* does an excellent job of surveying a broad territory while also performing readings of specific literary and historical texts.

Westphal's conception of a plausible world, while implicit throughout the book, is not discussed at any length in it. As noted, plausibility offers Westphal a helpful middle ground between the dubious idea of one true "real world" and the potential *mise en abyme* that the theory of possible worlds seems to invite. Plausibility makes possible the practical overcoming of the problem of referentiality, something Westphal found difficult but necessary in formulating his view of geocriticism in his earlier book; geocriticism, after all, relies on a great deal of poststructuralist theory, including notions—such as Jacques Derrida's notorious assertion that there is no *hors-du-texte*—that call literary referentiality into question. Certainly the Paris of Victor Hugo's *Notre-Dame de Paris* is not the same as the "real" Paris, since the former is obviously a literary setting in which fictional characters lead their fictional lives. But neither is Hugo's Paris *not* Paris, which must be equally obvious.[8] In *The Plausible World*, Westphal is able to consider simultaneously the places as they appear in literature and in "the world," while avoiding a position of either arrogance or ignorance. It is a bit like Gianni Vattimo's notion of *il pensiero debole* (or "weak thought"),[9] insofar as the *plausible* recognizes the humility with which the geocritic must approach the subject of geocritical inquiry.

This subject, I maintain, is the worldly world itself. Any geocriticism that is worth the effort must engage actively and consideratedly with the *worldly*, by which I do not simply mean the secular or mundane, although these are necessarily a part of it. I mean that geocriticism maintains a comportment toward the world that embraces the entirety of spatial and social relations, which in turn constitute the literary cartography produced in those multifarious ways of making sense of, or giving shape to, that world. Stated less circuitously, geocriticism approaches texts as literary maps that, regardless of the ostensible real or imagined spaces depicted, help us understand our world. Of course, by *worldly world* I do not mean to fall back onto some naïve vision of the "real" world, as opposed to figural spaces of fiction or of fantastic otherworlds. Indeed, I refer especially to that surprising and revelatory sense of Erich Auerbach's *irdische Welt*, which he discovered in the seemingly paradigmatic Otherworld of Dante's *Commedia*. As Auerbach writes, "the *Comedy* is a picture of earthly life. The human world in all its breadth and depth is gathered into the structure of the hereafter and there it stands: complete, unfalsified, yet encompassed in an eternal order; the confusion of earthly affairs is not concealed or attenuated or immaterialized, but preserved in full evidence and grounded in a plan which embraces it and raises it above all contingency."[10] Such a description would not be out of place in a discussion of the ways that literature serves to map the world, combining those material elements of earthly experience with the intelligible forms of the imagination. Not every map maker is the literary cartographer that Dante is, of course, but Auerbach's assessment of the Florentine "*als Dichter der irdischen Welt*" figures forth a critical approach to comparative literature and the worldly world in which live.

Ralph Manheim's translation of *irdisch* as "secular" highlights the irony of Auerbach's title, but it is not a literal translation; as in the quoted lines, *irdisch* means "earthly," whereas *weltlich* is closer to "secular" or "mundane." However, in this case, the earthly or worldly aptly registers the ways in which Dante's "divine" *Commedia*, even in its otherworldliness, truly represents the world as we experience it in life and literature, in all its literal and metaphorical worldliness. In this respect, the worldly world is also emphatically a plausible world. Geocriticism is necessarily *of* this world, and a geocritical approach to world literature may disclose novel representations and interpretations of its diverse, protean spaces.

Acknowledgments

I am very grateful to Christine de Buzon and to Martine Yvernault, who have been kind enough to read my manuscript. Claude Benoit and Domingo Pujante are also to be thanked for making possible my residency at the University of Valencia between January and July 2009, where I was able to devote myself to research and writing. And last but not least, let us not forget the Université de Limoges and the Limousin Region whose assistance has facilitated the publication of this book.

Some sections of the book have been borrowed, in revised form, from some of my articles and papers. They include the following: "Babylone au Moyen Âge ou la géocritique d'une trace," in *Babylone*, ed. Bernard Franco (Abou Dhabi, Paris Sorbonne Université at Abou Dhabi, November 2009); "Spazio, luogo, frontiera. Dante e l'orizzonte," keynote address, Congress of the Italian Association of Comparative Literature, Université de Cagliari, October 15–17, 2009; "Les horizons lointains d'Abou Bakari II. L'imaginaire atlantique à l'épreuve de l'imagination africaine," in the journal *Otrante*, edited by Kimé, "Mondes imaginaires," no. 24, 2008, pp. 103–16; "Quelques considérations sur une géocritique de l'espace africain," keynote address, APELA Congress, Université de Pau et des Pays de l'Adour, Bayonne, September 2009; "Cartographie et graphocratie ou de la pertinence des anagrammes. Michael Ondaatje, Nuruddin Farah, Kamila Shamsie," in *La production de l'étrangeté dans les littératures postcoloniales*, eds. Béatrice Bijon and Yves Clavaron (Paris: Honoré Champion, 2008), 37–47; "Atlas et catégories génériques. Une hypothèse de travail," in *Echinox* 16 (2008): 283–89.

Introduction

The world has long been one, or at least, it has long wanted to be one. And this unity was subsumed under a superior point of view that one attributed to a singular divinity, who is both solitary and insolate. The world is the ideal projection of the Occident, which quickly found its God and its genesis, or which has even found several rivaling versions thereof. For the ensemble of earths located there where the sun sets, the point was to legitimize their action (or rather that of their inventors) because each territory is the result of a human invention. The protection of higher and abstract beings was certainly comfortable, but the action remained nonetheless human. It was human to such an extent that it became inhuman, because that which is human can quickly become too human. However, it was necessary to arrange things so that these humans would not shoulder the entire responsibility of discrepancies in behavior, which were numerous from the start. Actually, the Occidental man has always known how to make do with his cowardice and his eminently plastic conscience so as to not have to completely own up to it himself. And it is in the name of an ideal harmony advocated and put into place by the great founding texts of which he is the author, and as a terminus ad quem, that this man has initiated the work of an ideological harmonization that is within planetary reach. We have known the resulting tragic consequences. The harmonization pretends to be salutary and splendorous, but it is unhealthy and nefast. It is transformed into a reductive and violent homologation. Beyond a shadow of a doubt, slaying[1] takes part in the great plans of the Occident, which has never been too bothered by the "collateral damages" it causes. One has lost sight of the harmony as a state, and that harmonization is a practice. And yet it is a practice that implies the mobilization of an energy that only asks to be released and to let go, becoming uncontrollable and delirious, restrictive at the very least. Before Theseus defeated him, Procrustes harmonized his world on a bed on which nothing was allowed to hang over the side. We are familiar with this very crude image of Greek mythology, which has entered into common vocabulary. The image has lost nothing of its relevance nor of its brutality.

The apex of this process is modernity, whose effort at homologation initially manifested itself in the enterprise of colonization, which deprived a part of humanity, frozen in an otherness without remedy, of its soul. The observation of the failure of modernity was late and scathing; it is without appeal. As pointed out by Jean-François Lyotard, modernity broke down *somewhere* in Poland in 1942. And the Occident with it. I subscribe wholly to this lucid judgment also pronounced by Adorno and others before him. Modernity is not, however, a ray whose line breaks down at some barbaric moment of history. It is endowed with a historical and maybe even spatial origin, even if it is poorly identified. Where could modernity have been inaugurated? In my opinion, *somewhere* in the Canary Islands. The old *locus amoenus* of the Greeks, for whom the archipelago sheltered the Macaronesian islands and the actual paradise of so many tourists, was also the testing ground for modernity. The experiences observed there were somber and foreshadowed even darker events to come. The mercenary troops led by Jean de Béthencourt took possession of Lanzarote in 1402 before other condottieri monopolized throughout the years the rest of the archipelago, which disappeared into the Spanish purse in 1496. The Guanchetos populated the Canaries during the prehistoric era. They were quickly reduced to slaves, massacred (sometimes even by mass drowning), or assimilated. In fact, they had just about been wiped off the map when Columbus reached land in "America." This island made up the test-bed of a modernity that would next affirm itself *somewhere* between Mexico and Peru. A little later, Africa and many other places on the planet became, in their turn, the theater of cruelty and of Occidental insatiability. The two chronic extremities of modernity coincide with the latest of human extremities. And that particular human accommodates himself poorly to humanists. The Occident is no stranger to paradoxes. The long parenthesis between this beginning and this end—is it worth a malediction? Nothing can exclude it, but to bear such a judgment is the privilege of great minds and of inveterate pessimists, among the ranks of which I do not count myself. However, one thing is certain: the Occident distinguishes itself through an articulation that makes of the world it draws the singular duplication of its own truth. The Occident is in the image of the gods who abhor rivals before them. And the universe, it preaches with an unfloundering conviction, is an extension of its own world. To sum up, as there exists only one God, there exists only one world. This founding idea is supposed to be obvious.

Had I been born a hundred years earlier, I might have heard Scarlett O'Hara exclaim with what little underscored enthusiasm she had left, facing the sight of the ruined fields she plowed, "Tomorrow is another day." The sun of the next day would rise on the same world on which the ashes would have barely cooled. But, in one way or another, the sun would rise. If it had been the case, it would have been easier to find a title for my book. Speaking of the title . . . Well let's take things in order. One thing is sure; it would not be called *The Origins of the World*. As it happens, more than one person considers this origin obscene, without doubt because it brings to mind the famous tableau of Gustave Courbet, painted a few months after Scarlett would have sighed with modesty from Atlanta. It would have been nice, however, to bring the first geography closer to the mountain of Venus rather than to Adam's coast.[2] We would have gained both in height and view, and the Eternal would have certainly been feminine. No, in such a marvelously uniform environment, this book would be called *The World*, without any more hassle. One would have just had some consideration for the ever-determining article, because *the* excludes "the one," which calls forth "the other." But after modernity, at the heart of a fumbling or already jabbering postmodernity, this title would have lost all its basis. The innocence of the world being forever lost, it would have been necessary to flush out an epithet that would qualify it. The simple determinant would not have been enough. Somewhere in Poland, while the world was being administered its last rites, the Occident had, indeed, become aware that it shared the same definite article, *the*, as grief, disaster, and uneasiness, from which it distinguishes itself less and less. After 1942, after 1945, after the mass diffusion of the news of horror, the excessively peremptory nature of a world declined in the singular popped up, even in front of the eyes of the self-righteous who didn't want to see or hear or understand it.

Postmodernity conceives of multiple worlds, an infinity of worlds destined to relativize the impact of a world on a unique model of modernity. All certitudes put aside, one felt they were "possible." One found the appropriate epithet: modernity no longer really engaged in anything. The cosmos gave way to more or less indistinct heterocosms. After others, and those who have gone before me (Thomas Pavel, Lubomir Doležel, etc.), I have had the chance to evoke the possible worlds theory. I will limit myself to recalling that, in the hypothesis of this heterocosmic proliferation, the world of reference—that one that encompasses the so-called objective real—would be a world among others that fiction created. In Wonderland, Alice evolves in a world that ontologically is contiguous to that which overlaps the White Rabbit's hole and where she is bored by her sister. One is the fruit of the imagination of Lewis Carroll, the other an "emanation" of Victorian England. Both are representations that maintain with the real a relation of degree that is undoubtedly different (so much so that different worlds are placed in a hierarchical relationship in regard to one another) but of an analog nature. Because, once one admits the hypothesis of the existence of a plurality of worlds, one also recognizes that *the* world is nonetheless relentless to a second-rate singularity to which it has become unrepresentable. *The* world of modernity, a support for the "objective" reality, is fragmented in a constellation of possible worlds whose representation constitutes, at best, an approximation. The possible worlds are postmodern. If they were associated to a new form of modernity—because for more than one critic (Jürgen Habermas, among others) postmodernity would be, *despite it all*, a late phase of modernity. These possible worlds throw back to the liquid modernity that Zygmunt Bauman had so often mentioned from the beginning of the new millennium. And as the great sociologist, who was himself one day obliged to flee to England from Nazism and his native Poland, remarked, "in a fluid and moving environment, the eternal truths remain ideas in the air."[3] It is sometimes preferable that the ideas remain up in the air, because there they can oxygenate themselves. And the air, like all elements, is in space. To talk about the world in the singular or the plural is to talk about space, about a space that maintains indissoluble links with time.

Back in the spring of 2007, when I had finished *Geocriticism: Real and Fictional Spaces*,[4] I let myself be cradled by a sweet illusion: that of having said, within the limits of my capacities, what I had to say about space and the modality of its esthetic representation. But the illusions were temporary. All it took was for me to have a change of scenery to discover not just any eternal truth but a sandy coast (shall we dare say the word: *beach*)—where one can stand still on the wet sand and let one's gaze wander, covertly sweeping the perplexity that has preoccupied generations of beach bum metaphysicians. What does the horizon close in on? On what does it open up? Is it an impenetrable frontier or a threshold waiting to be crossed? Definitively, what is going on with the status of the space? More than one navigator, and more than one female navigator, have taken the challenge to find within the ocean the beginning of an answer to the question that Isabelle Autissier, the first woman having completed a world tour in a sail boat all alone, has already summarized: "Can we resist the horizon? The naked line that delineates the ocean, that one sees from the shore, is a question, a promise, a call."[5] This interrogation has intrigued a few dreaming vacationers. It has cost many sailors their lives, starting with Dante's Ulysses, whose flame will illuminate a few pages of this book. Anyway, whether one be a dreamer or just too bored changes nothing in the affair, or very little. Because it is better to accommodate the assessment that, on the esthetic versant, will complete that of Isabelle Autissier: "One never finishes with space. One only ever speaks of it and within it. Never shall we leave it. To go where, I ask you."[6] Michel Serres asks this question in his nice essay on the painter Carpaccio.

To go where, indeed? And I will add to that the question mark that Serres omits, for a reason that escapes me, because this point encloses the minimal boundaries of the interrogation. The questioning is large and visceral. It is not just about a more-or-less abstract point situated along our existential path. It's the entire surface of our being in the world that it concerns. Because the space is that which, indefinitely open, deploys itself beyond immediate perception. It is not here; it is there, without being exactly beyond reach. Covered with a veil of intelligibility, it escapes first from human understanding. A desire that engenders excitation and chills is superimposed onto space, which is in itself always an enigmatic space. And the excitation is itself an incitement to mobility as we have understood from its Latin roots, because there are roots that are compatible with unorganized movement, as we have been taught by Gilles Deleuze and Félix Guattari, philosophers and botanists of knowledge. To confront space is to go forward to an encounter with an enigma, *elsewhere*, beyond the limits of the controllable territory. It's to leave in order to lift the veil that covers a mystery. This vision of space is the essence of our complex time. It has imposed itself rather late in a rather limited corner of the planet.

The chapters that follow are dedicated to a long-term geocritical investigation. There is a double purpose. First of all, we need to recontextualize the spatial vision of modernity within the mirror of its own history. If modernity is not coextensive of the Occidental past, this means that a passage took place at a given moment or during a determined lapse of time. It doesn't take a rocket scientist to figure out it took place around the Renaissance. Like others, I tend to place it so that it coincides with the first ultramarine installations of European powers. That takes us back to the indicative date of 1402, which announced that even more symbolic date, 1492. The entrance into modernity provoked a turn in the perception of space. With a firm steadiness, today one invokes the spatial turn that was produced at the heart of the postmodern era. It was identified by geographers, urban planners, sociologists, literary specialists, and many others. But, already during the sixteenth century, a discreet spatial turn had therefore modified the reading of the world, the old and the new. Cartography never stopped accompanying and illustrating the evolutions of this reading in the mentalities of the time. Less spectacular, almost underground, language and philology have brought their own contributions to the revival of the spatial turn by adapting terminology to emerging necessities. A powerful flow had led toward a novel representation of spaces, the intrinsic value of which one is finally starting to measure. If the postmodern *spatial turn* has enabled the rebalancing of the respective epistemologies of spatiality and of temporality, the modern turn has allowed the word *space* to leave its temporal dress in order to take on the meaning with which we associate it today. Space was, in the beginning, a space of time. I'll come back to that later, but yet another interest drives this investigation. To satisfy it, we will force ourselves to feed a line of thought that would contribute to ripping the reading of space from its traditional Occidental focus, to which, as I've just written, the universe of men is too often brought back. Combating a eurocentrism without nuances that—and let's be fair to the ancient world—has evolved over time toward Occidental-centrism supposes, as a matter of course, that one integrates into the reasoning points of view other than those developed by the West over the course of the centuries. In this book, we will talk about, among others, Aztec cartography, African and Chinese transoceanic navigation, and the spatial conception of Australian Aborigines. Without pretending to know in detail these alternate spatialities, I tried to take them into consideration with an Occidental point of view, whose perspective, with experience, seems to be less and less universal.

This trip across the ages and cultures will start at the center of the world, or more precisely, in its centers. The world disposes of so many centers, in fact, that it counts their declinations—and they are simply uncountable. The stopover in Greece will be longer than the others. At Delphi, through the medium of philosophers and ancient geographers, Zeus and Apollo delivered one of the most beautiful meditations on the belly button of the world, the radiating *omphalos*. Later, this geographic and metaphysical umbilicus inspired among other geographers the idea of an omphalos syndrome, which merits explanation here. The Middle Ages will be visited with diligence. By the way, the medieval maps—Christian and Muslim—have a lot to say about the orientation or the disorientation of a Western area that, from the beginning, was obsessed by the quest for a precarious balance. In the following chapter, we will move our gaze along the horizon line whose appeal, as Isabelle Autissier has so well understood, is irresistible. We will ask ourselves nevertheless if the horizon has forever exercised its fascination on dreamers and rowers. To do so, after having left Ulysses, we will accompany Dante on the beach of Purgatory to the antipodes of Jerusalem, which was the displaced center of medieval *Christianitas*. From this beach, the Florentine could have scrutinized the horizon. Did he do it? We will also survey the opening of the Middle Ages onto space and its curious manner of weaving together places. The third chapter multiplies the ways out beyond the porous limits of the European microcosm. After having witnessed the plunge of the diver of Paestum from the height of a Herculean column, we will follow in the noble wakes of Abu Bakari II, the Malinka emperor, and of Zheng He, Chinese admiral, who, like the Argonauts before them, chose to cross the horizon to discover what was hiding beyond. All these men, whose effigies should have left an impression on the world scene, have been pushed by a violent spatial urge.

They needed to let themselves go to the desire to know just a little bit more—or ideally they did, because their existence and their exploits cannot always be vouched for, but that does not matter. In a next chapter, we will witness the desperate efforts of Occidental man, struck by empty spaces, deployed to master the immensity of the lands and of the oceans that opened up before him. Facing this unbearable spectacle of a space that escaped his mastery, Occidental man did what no one other than himself would find necessary to do: to do his best to take control of this space with many reductions, restrictions, and manipulations destined to capture it in a *place*. The inveterate need to mold the open space into a closed place is one of the trademarks of the Occident. The history of this coercive mutation of space in place or in a conglomeration of places underlies the entirety of this present study. The partial end of this history occupies the last chapter where we will interrogate the destiny of the invisible lines and rings in water, on the colonial imposture relayed by cartography, on the intention to erase the Other—from maps, at least. To finish our itinerary on a lighter note, we will speculate (in the philosophical sense of the word) in the company of Giorgio Agamben, Peter Sloterdijk, Clément Rosset, and some others on the diss-measure that relativizes the so modern *métrise*,[7] or measured mastery, of the world. Because between two lines, at the detour of a few poorly joined points, there should be enough place left so that another world can rise. That world would integrate, as written by Nicole Lapierre, "the refusal of a binary conception . . . opposing fixed and essentialized identities, the attention given to the *routes* rather than to the *roots*, and finally the accent which is placed on the movement, displacement, more than on the territory and the establishment."[8]

But the title of this work still remains a question. It would be insufficient to not qualify the world. Qualifying the world as possible on the scale of all its history is vague. So? Maybe it is necessary to invest in the margins of liberty that subsist between the static and dull rigors of the singular world and the erratic engagement that is produced by the frequenting of the possible worlds. Perhaps between a surmounted singularity and an integrated plurality another world exists. That would be an oscillating world whose humility would display a certain elegance. This would just be a *plausible* world. Without pretension, it espouses the irregular form of the spatial puzzle that characterizes the planet, its history, and its current state. It goes without saying that a plausible world would sound the toll of hegemonic revindications of the West. They are, anyway, unlikely. Language often hides within the traces of its path subtle elements of derision. Therefore, the adjective *plausible*, which comes from the Latin *plausibilis*, shares the root *plaudere*, which means "applaud." Thus, that which is plausible is that which deserves to be applauded. The plausible world would therefore be a world that merits to be applauded. And why not? But is this applause ironic? And to say that this doubt is liminal . . .

CHAPTER 1

The Multiplication of Centers

The Location of the Omphalos

Truths often fall from the sky, just like ideas, flowerpots, or glass bottles. Let's take the case of a hunter bushman, a Sho, located in the middle of the Kalahari Desert, somewhere between Botswana and South Africa. Let's imagine that he found a soda bottle in the sand that a scruples-less aviator threw out his window into the void, a concept that is eminently relative. The first lesson of this fable, which is important, would be that the name of Occidental products is not universal, because the Sho, whose name is Xixo, believed he had found a gift that the gods made to his people. The message of the second lesson would be even more important. Xixo hurried himself to take the "bottle" to his own people, whose wisdom was put to the test. What purpose could this curious meteorite serve? How should they interpret this divine gift? As, in the end, doubt about the object created tensions at the heart of the Sho community, Xixo was asked by the council of his village to travel to the world's end in order to get rid of the encumbering thing: "We don't want the thing. You should get rid of it yourself." In short, Xixo was very disappointed. He felt that it was being disloyal to the gods to require him to throw the thing out of the world. As a result, he started to ask himself if the gods really existed. For Xixo, the world's end was not so far. Along the way, he had the chance to live some adventures, as told by Jamie Uys in *The Gods Must Be Crazy*, a South African film that, in 1980, flew Botswana flags in movie theaters around the world in order to circumvent restrictive apartheid laws. The portrait of the Sho as good savages living in a utopian space would have surprised the wisest of viewers. Xixo and his people did not live outside of the "real" world; they shared the nationality and the troubled period of the director, and his injustices as well. But in any event, the tale of Jamie Uys is inscribed in a tradition that the Occident contributed to establishing.

On the horizon in Greece, the vast Uranus, heavy as a thunderous sky, had the habit of weighing down on Gaia, the maternal earth, and pushing back into her entrails the children that she conceived from his works. So, reveals Hesiod in his *Theogony*, Gaia encouraged her son Cronus to castrate Uranus using a sickle of flint. With Uranus's retreat, enough space was created between the earth and the sky for life to see the light of day. Cronus succeeded in his mission. Then he married Rhea, who was his sister. But Cronus became just as weary as Uranus, because his parents had confided in him a secret: he too would be dethroned by one of his offspring. He therefore undertook the act of eating his children. Hestia, Demeter, Hera, Hades, and Poseidon were all ingurgitated before Rhea had the presence of mind to save Zeus, her sixth child, by giving his ogre of a father a stone wrapped in a tongue to eat. This tongue constituted the first draft of a language of fiction. The appearance and the essence of things had stopped coinciding. The tongue instituted a signifying and significant gap between the referent (the child) and its representation (the swaddled stone). This occurrence was the very first counterfeit. This original simulacrum enabled Zeus to find his place on the Earth. In order to protect the child from the revenge of Cronus, Rhea took the precaution of giving him over to Gaia. Gaia kept him in a cave at Mont Ida, in Crete, where he was raised by the Curetes. The rest of the story is well known and too long to retell here. I believe we would barely be stretching the link by putting together the prodigious rescue of the future master of Olympus from an ancient emergence with a "narrative of the earth." Zeus owes his life to Rhea's invention, which had been suggested by Gaia, both of whom are mothers and earth, matrices of all imitation, of all representation. As for *geô-graphia*, it is the narrative that will describe the body of Gaia, identified at the surface of the world, at the original space.

Later, as told by Hesiod, Zeus married Metis of Oceanid, who told him how to get Cronus to give back his five brothers and sisters. Cronus vomited up Zeus's siblings, and Zeus took advantage of this to recover Rhea's stone, his own simulacrum. When Metis was pregnant, he got in a hurry to swallow her whole, because he too doubted his descendants. (Cannibalism often tints the beginnings of Greek culture with blood.) A little later, Zeus came down with a bad migraine. Hermes, who noticed it, sent Hephaestus to open up his cranium with a stroke of a mallet. Athena, all armed, came out through the crack. But we aren't going to follow her to Athens; we are rather going to take the path of Delphi, where the *omphalos* is located. When it came time to determine where the center of the universe that he reigned over was located, did Zeus remember the stone that saved him from Cronus's gluttony? According to Pindar, he released two eagles from the two extremities of the horizon (Oriental and Occidental) to establish the geometric center. The two birds of prey crossed paths above Delphi, in Phocis. He concluded that it must be there that he should locate the navel of the world, the *omphalos*. This location was marked by a convex white stone that was wrapped in ribbons and covered in a large mesh netting. In memory of the first winged birds of prey, the *omphalos* was flanked by two golden eagles, as indicated by Pindar in his fourth *Pythique*.

The location of the *omphalos* has always been designated according to "scientific" criteria, without regard for the sacred hierarchy of space. And it's too bad if a skeptical (or shrewd) mind objects that it's the choice of the place that determines, a posteriori, the pseudo-objectivity of the measure. Was it the same stone that was used by Rhea to trick Cronus? Or, was it a different one, as Pausanias thought? The opinions of Greek writers diverge on this point, as on many others. At the end of the nineteenth century, cutting to the heart of the matter, August Bouché-Leclercq esteemed that the *omphalos* was a recent copy of Cronus's stone that had replaced the original "when the tradition of the 'omphalos' became credible. The 'small' egg-shaped stone [Cronus's stone] was first and foremost mobile: the geodesic center of the world needed to be indicated by a more robust marker, implanted in the soil."[1] All in all, the small mobile stone gave way to a large immobile stone, just like sedentary peoples superseded nomads and cities made of stone and marble replaced makeshift cities. The *omphalos* translated the desire that the Hellenic people had to have access to a uniform, stable space, which was a priori central. In short, Greece went about establishing a place. The *omphalos* was the center of a civilization that presumed itself to be the agent responsible for distinguishing the center of the world. On many different coins that have been found, the *omphalos* was a point indicating the center of a circle that occupied the surface of the piece. This vision was the result of a considerable evolution because, at the beginning, the *omphalos*, set within an exclusively sacred vertical perspective, corresponded to the spatial crystallization of a link between the sky and the earth. In many diverse cosmogonic narratives, like Genesis, this link is expressed in a language of stones, by towers or pyramids erected by architects, which were more or less real, more or less mythical, such as the tower of Babel, built in the image of Nimrod. If, at Delphi, the *omphalos* was at first included in the vertical vision (earth-sky), it ended up by being integrated in horizontal projections (earth-sea) that established the centrality of Delphi and, a fortiori, the Hellenic Ecumene, which was deployed around the Archipelagos, or in other words, the Aigaion Pelagos, the Aegean Sea, the Archi-Sea.

When looking through a dictionary of Greek mythology, there are two entries that follow each other closely: one is about Omphale and the other Omphalos. Omphale was queen of Lydia, famous for having enthralled Heracles. She did in fact reduce the half god to a slave, and then she married him after discovering his true identity. A curious inversion of roles took place next within the couple, at least in the later version of the myth, in the Roman era. While Heracles cross-dressed as a docile spouse tied to the domestic sphere, Omphale brandished a wooden club, the traditional attribute of her husband. This story is well known, because the arts have favored Omphale. In Heracles's company, she adorns an updated mosaic in the charming village of Llíria, not far from Valencia. The couple also takes part in paintings by Cranach le Jeune, Rubens or de Tischbein, a symphonic poem of Saint-Saëns, and even Italian togas. We might ask what relationship the beautiful Omphale has to the Omphalos of Delphi. It may be a pure onomastic coincidence. To be honest, this easy answer is quite unlikely, because the Greek mythos is the archetypical text: it tirelessly weaves links between all histories of the world. Nothing is left to chance. Therefore, it is not surprising that Apollodorus suggested a plausible explanation. Prior to contracting a mysterious disease, Heracles killed Iphitos. He went to Delphi to purify himself of his crime and his illness, the latter being the consequence of the prior. Seeing his initiative fail, he pillaged the temple and stole the enigmatic Pythia's tripod, in an effort to create his own oracle. To avoid bloodshed between Apollo (who was enraged by this episode) and Heracles, Zeus ordered the half god to accept three years of slavery in Lydia, at the end of which he would be cured of his illness. The man with the wooden club obeyed. Maybe Omphale herself embodied a fleeting desire of sedentary lifestyle on the part of the impenitent nomad Heracles. Maybe, subjected to this seduction that turned him away from his frenetic wandering, from his permanent mobilization, Heracles decided to explore in Lydia (the land of Omphale, the alternative *omphalos*, the zero point of space and of the traditional sex) another type of sexuality in order to confer a different and new form to his visceral need for a quest.

One thing seems sure: the *omphalos*, the object, the materialization of the center of the world, was located in the adytum of a temple that Apollo had erected in the memory of Python, whom he had killed. Pausanias claims, however, that the *omphalos* is located next to the temple. For him, it was as if the center of the world was slightly shifted. In any event, the *omphalos* could not be far from the tripod from which Pythia rendered her ambiguous oracles. The closer one got to the center, the less clear the words were, and the more fateful she was. But was the *omphalos* really the center of the world? In the *Odyssey*, a strange verse, the fiftieth of the initial song, instills a doubt. When it comes time to qualify Ogygia, Calypso's island, which is the furthest of those investigated by Ulysses, Homer notes that it is the *omphàlos thalássēs*, the "navel of the sea." So, like Varron (Caesar's librarian) in *La langue latine*, there is a play on words, and we can say that the navel of the earth is not necessarily located at its middle; we could even add, just as he did, that "Delphi is not placed at the center of the Earth, and the navel is not placed either at the middle of the human body."[2] At the middle of the body, he reminds those who might have forgotten, there is that which is hidden: the genitals. All things considered, the *omphalos* of Delphi was the tomb of Python and nothing more. Varron was quite prosaic. I would like to think that for Homer—in any case for Ulysses—the center of the world was located at its outer limit. In this way, there would always remain a half for him to discover, the Other. Still, it would be necessary to conceive of this point "beyond" knowledge. That's what the Homeric Ulysses undertook. That's what the Dantean Ulysses continued to do. That's what we try to do when we get it into our minds to cross the horizon line.

In the first lines of *On the Cessation of Oracles*, Plutarch narrates an intriguing anecdote. Epimenides of Phaistos, a wise man according to some, a prophet according to others, interrogated the oracle of Delphi on the validity of the myth and the reality of the flight of the eagles. Pythia, so the story goes, gave him an evasive and ambiguous answer. Epimenides then said, "There is no umbril of the land or sea: God only knows, man knows not, if there be."[3] The Cretan Epimenides was just as imprudent as Dante's Ulysses. He abstained from taking a boat to go verify the measurements, but his ideas reached Apollo who punished him for having asked Pythia such an offhanded question. For the god, it was understood that the *omphalos* could not be assessed "by the touch."[4] Many elements make it possible to envisage the hypothesis according to which the consecration of the central navel (which references the horizontal line of the world) was produced at the precise moment when the Greeks stopped having an absolute faith in their gods and in their myths.

"Did the Greeks believe in their myths?" asked Paul Veyne, in the title of a famous essay, which appeared in 1983. Yes, certainly, but before they made Delphi the center of the world. That is what, in any case, the discrete hesitations allow us to think, as, little by little, the different layers of the myth superimpose on themselves, introducing a spirit of play into the interpretation. Either one is just a little to the side of the center (Pausanias) or one downright deports its (Homer); one denies the validity of the metaphor of the center (Varron) or one even pushes to the side the possibility of such a myth (Plutarch). However, nobody incited Epimenides, as one day Xixo, to go throw away the *omphalos* in the great beyond of the world, or to throw it in the waters of the River Okeanos, which girded the known universe. Around Greece and the Mediterranean, reasoning continued to be based on the center of the world, the centrality of a point of view, of centrism, and soon eurocentrism. The West was already in the process of building itself. Plutarch, who had reported the episode, hurried himself to point out that in his time, the first century of our era and the beginning of the second century, the centrality of Delphi had been confirmed. Demetrius the grammarian left Britain where he had led an exploration mission to return to Tarsus, in Cilicia. For his part, Cleombrotus the Lacedaemonian had left the banks of the Red Sea, where he too was traveling, to return to Greece. The two men were coming back from the two extremities of the world, and they just happened to cross each other, by chance, in Delphi. Like the former eagles, they had recourse to a lapse of time equivalent to that necessary for covering the halfway distance that ran from the end of the world to its center. As for the sanctuary, it definitively ended its function at the end of the fourth century, when the Emperor Theodosius, a Christian, forbade "pagan" cults. The gods had become demons to whom it was best to avoid listening.

The Omphalos Syndrome

Delphi marked, so one said, the point of departure for all the roads that crisscrossed Greece. Every city, and a fortiori, every country possesses today an analogous landmark. It was Rome that set the tone. If all roads led to the Eternal City, it was equally necessary that they departed from there as well. Around the year 20 BC, Augustus had a *Milliarum Aureum*, a Roman golden-stone mile marker, built near the temple of Saturn. In fair play, Saturn, the Roman version of Cronus, recovered that which Zeus/Jupiter made him vomit: a big stone. Rome had taken the place of Delphi. The position of an *Umbilicus urbis Romae* has been calculated, between the Æmilia and Julia basilicas, at the heart of the Forum. Today, this *umbilicus* has miserably survived inside brick turrets. *Sic transit gloria mundi*... A few years ago, on the night of the victory of his team against that of Lazio, Francesco Totti, star of the Roma team, sealed the fate of the ancient empire's capital: "To win the *scudetto* (the Italian soccer championship), we have to beat provincial teams like that of Lazio." No more, no less. Rome, the capital of a provincial Latium? The charismatic soccer player of the Roma team wasn't even kidding.

A detailed study devoted to these symbolic landmarks would certainly attract attention. The zero point of Paris is located on the Notre-Dame parvis. It is also "the kilometer zero for the roads of France," as indicated by the plaque that marks it. The latter is, however, more recent than the Michelin maps, which, between 1911 and 1913, offered the first hierarchical numbering of the "roads of France." In Spain, the choice of the center was based on the Puerta del Sol in Madrid. In 2002, the plaque indicating the *origen de las carreteras radiales* (the origin of the big roads) had to be turned 180° because the map of Spain for which it served as an indicator had been wrongly oriented—it had been upside down. Maybe it is for this reason that in a Spanish comedy by Yolanda García Serrano and de Juan Luis Iborra, *Km.0* (2000), the 14 disoriented characters, who have a rendezvous at Puerta del Sol in Madrid, establish unexpected relationships and plan to set off on a new foot. And it is perhaps also this center that inspired the images of a refrain by Ismael Serrano: "Kilometer zero, in the center of a city breathes the soul which distances itself and escapes. Kilometer zero, the beginning of days to come, of the calm which will bring the storm."[5]

In India, the zero kilometer corresponds to the location of Mahatma Gandhi's tomb in Delhi. In Budapest, however, several sculptures have succeeded each other over the course of the years and different styles: the brave socialist worker, built in 1953, was replaced in 1975 by a stone zero, evoking the form of an egg and moreover situated at the entry of the tunnel that runs the length of the Széchenyi chain bridge, on the River Buda. In the United Kingdom, it's at Charing Cross, which is known for its bookstores, that the official Zero Milestone is located. This landmark has some serious competition, though: a stone sheltered behind a grill, in Cannon Street. According to the legend, it is the *omphalos* of London, the placing of which is attributed to the mythic founder of the city, Brutus of Troy.[6] The niche where the stone has been conserved for a while has been overcome by a series of billboards, singing the praises of diverse brands of sports shoes.

Let's take things a little farther. To Jerusalem, for example. In the domain of cartography, Jerusalem became a stable landmark from the time that the Christian world was represented. For Isidore, Bishop of Seville at the beginning of the seventh century and author of *Étymologies*, which was a sort of database ahead of its time, "in the middle of Judea we find the city of Jerusalem, which is a type of navel for the entire region (*quasi umbilicus regionis totius*)."[7] It is not very clear if, for Isidore, Jerusalem was situated in the center of Judea (of which it could have been the navel), or if Jerusalem occupied the center of the world, even if the first term of the alternative seems philologically the most likely. As it is sometimes easier to draw a diagram to make oneself understood, Isidore attached to his text a map that would know many imitations: the map known as the "T-O." This map draws its inspiration from ancient Greek models that place the disk of the world in the middle of an immense ocean. The world that Isidore's map reproduces was structured according to a Christian anagrammatic logic. The *T* that formed the rivers separating the three known continents represents the cross, the two bars of which joined together at a point that was obviously the center of the circle and going out from the center of a world that was encircled by the River Okeanos, the *O* by antonomasia. But this point was anonymous. While it seems to correspond to Jerusalem, it was not named. The text of *Étymologies*, just like its cartographic illustration, indicated an ambivalent *omphalos*, the center of a region, the center of a circle and, *maybe*, the center of the *orbis terrarum*. Isidore died in 636, two years before the Arabs seized Jerusalem. The model developed by Isidore remained a reference for some time. The East continued to occupy the top of maps, and the principle of dividing the continents into three (tripartition) continued to order the world. Circularity was not completely imposed. At the end of the eighth century, another Spanish man, Beatus of Liébana, brought a rectangular form to the map based on a commentary in the *Apocalypse* of Saint John. He decided to plot out the holy places mentioned in the Scriptures on the map. But whether they be circular or rectangular, these first mappemondes tended to put the role of Jerusalem into perspective. Beatus of Liébana, an Asturian monk, signed his named to these comments, which were based on the maps, and the original Beatus was a model for other Beatuses, such as, for example, the Gerona Beatus, at the end of the tenth century, or Leon Beatus, commissioned by King Fernand I of Castile and Queen Sanche in the eleventh century. All these precious documents gave even more importance to the position of Heaven than to the position of Jerusalem. Adam and Eve, flanked by the inevitable serpent, sat enthroned on their summit, but, even if they were very visible, they were mainly just on the right margin of the map, red, like the sea of the same name, while the South continued to occupy lots of space, which for us would be the East.

Toward the thirteenth century, Jerusalem ended up acquiring the status of *omphalos* as declared by the Christian universe and its graphic representations. This late conclusion is easily enough explained. Using as a pretext the end of the inadmissibility of the Seljuks' opposition to the Christian pilgrims who wished to render themselves at the Saint-Sepulcher, Pope Urban II launched, in 1095, from Clermont-Ferrand, the first "crusade," which did not yet bear that name. Four years later, this crusade led to the seizing of Jerusalem by Godefroy of Bouillon's troops. The city would remain in Rome's orbit until it was reconquered by Saladin, in 1187. Besides the brief period between 1229 and 1244, the city escaped from the grip of the crusades, which ended by drawing all attention toward Constantinople. In other words, the center of the Christian world was placed out of reach of those who had attributed it based on faith in a text that they believed to be sacred. John had previously conferred a celestial dimension onto Jerusalem. In chapter 21 of the Apocalypse, he evokes the apparition of a new Jerusalem after the victory of God over Satan, at the end of time. This shimmering city herself possesses a center, which subsumes under it the Lamb of God. Jerusalem is unstable, floating somewhere between heaven and the earth, and this indecision anchors it not without realism in the unstable reality of the period.[8] A vague and sacred place, Jerusalem formed both the risk of a future conquest and the promise of a beautiful dream, which would be best able to renew day after day, night after night.

The situation was not without interest. In 1987, in a study on the heritage of Ptolemaic geography in the Renaissance, the art historian Samuel Y. Edgerton employed a pithy formula, "the Omphalos Syndrome," which Anglo-Saxon criticism cites each time now that it associates cartography and cultural navel-gazing. The Omphalos Syndrome, according to Edgerton, in fact affects all "where a people believe themselves divinely appointed to the center of the universe, shows its symptoms in the history of cartography as often as in ancient city planning."[9] To cite an example, he mentions Ebstorf's mappemonde, developed in a Benedictine convent in North Germany, under the aegis of the Provost Gervase of Tilbury. This map of the world, which dates from the years 1220–30, spanned the centuries sitting on a forgotten shelf in the Ebstorf monastery, which was later transformed into a convent. A particularly meticulous Benedictine Sister undertook the task of dusting off the shelf in 1830. She discovered the world map and understood its importance. In 1943, the precious document was destroyed in a bombardment, but as there exist a few facsimiles, it is not completely erased from the collective memory. On the contrary, the Ebstorf world map is one of the most famous maps of the Middle Ages. Its appearance bears many resemblances to that of numerous shrouds that, from here and there, would have covered Christ's remains. It's as if Christ appeared superimposed: his head crowns the summit of the map, as it should, except for in the thirteenth century, the top often continued to indicate the east. A hand passes to the left (the north), the other to the right (the south), and the feet—as we could have suspected—are on the bottom, to the west. This world map disposes of an expected *omphalos*, Jerusalem, which is symbolized by the sacred heart of Christ. It is not an isolated example. During the same era, a Pantocrator (all powerful) Christ, which is today conserved in the treasury of the cathedral in Toledo, decorated the superb *Biblia de San Luis*, which Blanche of Castile had designated for the education of the future Saint Louis, her son. This Christ holds the world at arms' length, in an appeasing gesture, as if the *omphalos*, located at the level of his navel, guaranteed the balance of a universe whose followers would have acquired the mastery. Just as Edgerton remarks to a distracted reader, the Christian *omphalos* remained in the hands of the Muslims at the moment when the majority of world maps were conceived. The *omphalos* forms then a desired center that was unpositionable and inaccessible. Consequently, maps form powerful propaganda tools that the promoters of the different crusades knew how to exploit.

In 1964, the very young Giorgio Agamben played the part of the apostle Philip in the *Evangile selon saint Matthieu*. Pier Paolo Pasolini shot the film in a nearly secret fashion, with the support of just a few close friends and family members: his mom (Mary), but also certain writers, such as Enzo Siciliano (Simon Peter), Natalia Ginzburg (Mary of Bethany), Juan Rodolfo Wilcock (Caiaphas), and Alfonso Gatto (Andrew). Everyone feared having to endure the wrath of the Vatican, because, in March 1963, Pasolini was judged for violating the religion of the state and condemned to four months of imprisonment. He was criticized for portraying, in *Ro.Go.Pa.G* (a film of sketches cosigned with Rossellini, Godard, and Gregoretti), the crucifixion of a poor beggar from Borgate, the popular neighborhoods on Rome's periphery. The character died on the cross from indigestion of ricotta cheese. The tone of *Evangile selon saint Matthieu* was extremely different. Jerusalem took center stage, of course. But this

Jerusalem, taking into consideration Pasolini's budget, had been transplanted to the South of Italy, between Puilles, Basilicate, Calabre, and Rome. Pasolini's film was not sanctioned by the Vatican. It even drew praise for its author. As for Giorgio Agamben, he defended his *tesi di laurea* on Simon Weil the following year. In 1977, he published the essay that made him known beyond the borders of Italy: *Stanze. Parole et fantasme dans la culture occidentale*. It had nothing to do with Jerusalem but was about melancholy, the Pygmalion myth, and fetishism. The concept of the stanza can nevertheless help us explain what Jerusalem represented in the perspective of the Christian West, at the beginning of the era of crusades.

According to Agamben, for the poets of the thirteenth century, the stanza was the essential core of poetry, the foyer of *joi d'amor*. It was destined to remain out of reach. In an effort to explain the unreachable character of the stanza, Agamben took inspiration from a few passages of Plato's *Timaeus* and Aristotle's *Physics*, which cites *Timaeus*. At the beginning of book 4 of *Physics*, Aristotle undertakes a study of place and observes, "The physicist must have a knowledge of Place, too, as well as of the infinite—namely, whether there is such a thing or not, and the manner of its existence and what it is."[10] We can draw many lessons from this lead-in. It teaches us, for example, that the place and infinity do not share the same nature. In the following paragraph, Aristotle continues, "All suppose that things which exist are somewhere (the non-existent is nowhere—where is the goat-stag or the sphinx?)."[11] We can see that they are present on maps and portolans of the Middle Ages and the Renaissance, which Aristotle could not foresee. But, for the moment, let us content ourselves to affirm with him that there are beings and nonbeings that, like the goat-stag and the sphinx, only exist in language and therefore, sometimes, in our imagination. And Agamben remarked that neither could be found "anywhere, without a doubt, but maybe because they are both *topoi*."[12] It must be the same for Jerusalem, a singular topos escaping from the purely material spatiality of the eyes of those for whom it is an inaccessible *omphalos*. And this "place," as any place taken in terms of topos, is "like a pure difference, however endowed with the power to do as such so that 'that which isn't, in a certain sense, is and inversely, that which is, in a certain sense, is not.'"[13] For cartographers, as for the poets of the Middle Ages, it would be about applying the precept that Agamben sums up in this way: "it is not possible to appropriate the real and the positive without entering into contact with the unreal and the unlikely."[14] Jerusalem went about occupying, in a systematic and especially *explicit* way, the center of medieval maps and world maps until the time when the use of the term *to meet* was generalized, around the thirteenth century. Indeed it was in 1174 when the Anglo-Normand author Guernes of Pont-Sainte-Maxence forged the expression *soi cruisier*[15] in his work *Vie de saint Thomas le Martyr*, devoted to Thomas Becket, the Bishop of Canterbury assassinated in 1170.

Eventually, the dimension of the unreal stopped being an integral part of the mental and political landscape of the West. The vague desires of making the fantasy come true by weapons and violence would end up imposing themselves. The stanza would never have stayed empty for so long without nourishing melancholy, an ever so European humor. That being said, a part of the unreal was called forth to remain because it was necessary to continue feeding the dream of a realization to come. At the moment when it became possible to "meet" and to make of Jerusalem a nominal *omphalos* of mapped *Christianitas*, a Christian appendix is projected beyond the Muslim world, in a terrestrial *beyond*. In a chronicle dating from 1145, the Bishop Otto von Freising mentioned for the first time the existence of a Christian monarch, whose Kingdom extended beyond the Muslim world, near India. This is how the legend of Prester John, a Nestorian and a heretic, was born, which fed many medieval lies and military fantasies. Twenty years later, a missive that Prester John was supposed to have sent to Manuel I Comnenus of Byzantium, and that in fact was a counterfeit, started to circulate in Europe. In 1177, Pope Alexander III felt obligated to send an emissary to Asia, charged with the mission of responding to Prester John. But something bad happened: there was no news from either the king or the emissary. We could spend many long pages on this extraordinary legend, maintained by Marco Polo, Mandeville, and Joinville. Those who are really interested by the story can always read *Baudolino* (2000) by Umberto Eco.

Giorgio Agamben recalls in his essay that the "stance" (the stanza) is an Italian transposition[16] of the Arab word *bayt*, which means "residence." The stanza is also a composition that is articulated around a principle line where the object of desire expresses itself. What Agamben does not explain is that the *bayt* is no kind of residence. It was initially a sacred tent in which idols were kept—the Arab *Bayt î lis* like the Jewish *Beith-el*. Both come from the word *bétyle*,[17] which indicates both an *omphalos* and a meteorite—that is to say, an object that has fallen from the sky. According to the legend, the *Bayt îl* originally held the stone or the pearl offered by God to Adam. In the Koran, Abraham and Ishmael make of it a worship site, which is at the origins of the Ka'ba, in Mecca (4:125). As for the *Beith-el*, it is the stone that served as a nightstand for Jacob during his dream of the ladder: as the son of Isaac understood that this stone indicated the house of God and the door to heaven, he straightened it up, and anointed it with oil at the summit to make of it an *omphalos*. This took place in Bethel, in Samaria (Genesis 28:11–19). Today, the small town of Beitin occupies the likely site of Bethel. It is located five kilometers from Ramallah, on the West Bank of Jordan. In 1977, an Israeli colony was implanted not far from there under the name of Beith El.

Orientations

Under the reign of Roger II, King of Sicily from 1130 to 1154, the cartographer Al-Idrissi, a native of Ceuta, drew a map of the world that can be considered among the most spectacular in history. It is striking in more ways than one, starting with the choice of colors. As Predrag Matvejević, one of the contemporary bards of the Mediterranean, notes, Al-Idrissi gave great importance to the chromatic palette: "He took up Ptolemy's climatic studies and attributed to each one of them its own color: the Mediterranean circle and its 'fourth climate', in which the green and blue of the sea dominate, the yellow of the dessert, the red, from the most blazing to the darkest, that of the rising and setting of the sun on the sea or the desert. He gave the Atlantic Ocean dark colors: the Arabs call it the Sea of Shadows (*Bahr al-Zulumat*)."[18] Al-Idrissi's *Livre de Roger* (1154) contained a large silvered world map, which was unfortunately destroyed in 1160 during a riot. Today, a more modest map remains, which gives us an idea of the cartographer's immense talent. This map is oriented neither to the east nor to the north, but to the south.

The center of the circular (or rather sphere-like) world of Al-Idrissi is located in the Arab Peninsula. The position of Mecca is not emphasized. Undoubtedly it is obvious. *Omphalos* by antonomasia, the Ka'ba sheltered by Mecca is the place toward which all believers orient themselves at the time of prayer.[19] The map here takes on a particular importance, because religion and geographic orientation are closely linked. As for the sources supposedly coming from the Nile, a little further up, toward the right (I wouldn't dare speak of the northeast, which would be the southwest in the modern perspective), they are quite visible. The southern orientation of the map perturbs the habits of the Occidental observer, who, since childhood, is used to seeing the planispheres in a certain direction—*his or her own*. In the spring of 2009, concerned about handling visitors' cervical vertebrae with care, the organizers of an exhibit dedicated to cartography in Requena, in the up-country of Valencia, preferred to turn the reproduction of Al-Idrissi's masterpiece in the "right" direction. At first they pivoted it to the north. I suppose though that the Arab-speaking visitors, for their part, took the risk of a having a crick in the neck while trying to read the legends of the map. This little game of "orienting" the world in function of a particular *omphalos* later inspired an Australian apprentice cartographer.

Weary of always hearing that his country is located at the bottom of the map, Stuart McArthur decided, at the age of 12, to inverse the usual perspective and to place Australia to the north, orienting his world map to the south. It would seem that a quite traditionally oriented geography professor, to whom the young student submitted the project, rejected it, considering that it was "incorrect." In 1979, when he was 21 and of majority age, Stuart McArthur produced, in collaboration with the University of Melbourne, the *McArthur's Universal Corrective Map of the World*, in which Australia crowns, literally, the world. This map had outstanding success. The first copy of it that I saw decorates a wall at the library of Texas Tech University, in Lubbock, in West Texas. I might as well confess that my first reflex was to turn my head to the side to put the world back into its proper place—but which one was that? McArthur's undertaking was amusing, but it was not so original. As early as 1566, Nicolas Desliens, a great cartographer from the Dieppe School, produced an "inverted" representation of the world on a planisphere produced on vellum. In this way, he attributed a salient place to the *Terra Australis*, which had not yet been discovered.

For a long time, the existence of this *Terra Australis* was imagined. Ptolemy had posed the hypothesis, as had other authors of various *Beatus* maps, who added southern antipodes to the tricontinental representation of the T-O maps. It was necessary to find a way to balance out the other side of the earth, in the antipodes. To do so, many scholars of the twelfth century opted for an *omphalos* that was relocated to the south, in the *terra nigrorum*, according to the terminology of the period. This other navel of the world was Arin, which is still called Aren or Aryn.[20] All the scholars who defended this theory happened to be living or working in Spain or in Sicily, which meant having contact with the Arab culture. Moshé Sefardi, who was later known as Pierre Alphonse (Pedro Alfonso), to whom the creation of the toponym is attributed, was a native of Huesca, whereas Plato of Tivoli, an Italian translator, officiated in Barcelona, and the Englishman Adelard de Bath had stayed in Spain (or in Sicily) before following Pierre Alphonse. The following century, in his *Imago Mundi*, Gossuin de Metz mentions, for his part, a "city which has the name of Aaron. It is situated in the middle of the world, and is made all round."[21] Gossuin's Aaron is certainly Pierre Alphonse's Aren,[22] which Roger Bacon associated next with Syene, which is also called Aswan, the southernmost city of Egypt. Geography sometimes resembles a Rubik's Cube. The story goes that it is at Syene that Eratosthenes tried to determine the circumference of the earth because of the proximity of the Tropic of Cancer. The desire for balance brought on this type of speculation. Arin is, in any case, a nice example of an *omphalos* that brings Greek, Muslim, and Christian cultures closer together.[23]

However, it must be recognized that this *omphalos* did not have great success . . . Even though . . . let's change latitude; let's go to 53° south. In *Horizon mobile* (2009), novelist Daniele Del Giudice tells of three voyages around Punta Arenas, Chili, and the Argentinean Tierra del Fuego: that of the Italian Giacomo Bove in 1882, of the Belgian Adrien de Gerlache between 1897 and 1899, and his own, in 1990. During the last week of the southern summer, Del Giudice arrived at the Centro Geográfico de Chile, at the very southern part of the country, somewhere on the Strait of Magellan, at Port Famine. The Centro Geográfico de Chile is not a research institute dedicated to the geographic or geological study of cold zones but a cippus of white stone, a zero kilometer marker erected in a deserted landscape. What an odd center . . . what a strange *omphalos*! The stupefaction of Daniele Del Giudice can be heard: "But how can it be that I find myself at the very edge of Chile?"[24] The answer is as simple as it is surprising: "The fact is that the country has covered the Drake Passage and it has annexed a portion of Antarctica, a piece between the 82nd and the 53rd meridian which closes in on the pole. Argentina and Great Britain claim this territory but Chile has already carved out the same four thousand kilometers between here and the Northern border with Peru. That is why I am at the 'center.'"[25] The center, therefore, isn't Santiago, but Port Famine, in precarious balance between the official Chile and a Chile that is just as virtual as the antipodes of Pedro Alfonso and Gossuin. Port Famine was previously known as "Rey Don Felipe." The explorer Pedro Sarmiento de Gamboa determined the location of the colony in 1584. Three years went by before the privateer Thomas Cavendish happened along. He discovered "the rest of the people dead from cold and starvation."[26] Rey Don Felipe was then rebaptized as Port Famine (Puerto del Hambre). A few centuries later, Port Famine/Puerto del Hambre became the center of Chile. There, Pinochet confined a few of Salvador Allende's ministers; so it can occur that one is confined to the center.

The signification of this alternative center is very different to the Chileans and to Pierre Alphonse. For the former, a center placed in an apparent margin is a revendication. For the latter, it could be that hiding behind Arin or Aren there would be another center, which is even more secret. Indeed, I wonder if Pierre Alphonse did not make of Aren the center of a very intimate world: his own. Still today, there exists in the province of Huesca a charming little village known by the name of . . . Arén. Could an exotic Aren hide a domestic Arén? *E se non è vero è ben trovato*. In that case, Pierre Alphonse would have had the same sense of humor as the German orientalist Friedrich Rückert, who, in the nineteenth century, considered that the center of the world was Germany; the center of Germany was Franconia; the center of Franconia was Schweinfurt; and the center of Schweinfurt was his house; and the center of his house was the heart of his beloved. Sometimes, the *omphalos* is private.

Sometimes, organizers of cartographic expositions let themselves go to the temptation of reestablishing the *order* of the world. Moreover, it was not until the eighteenth century that the Nordic tropism was officially consecrated. In Europe, throughout the entire Middle Ages, world maps continued to be turned either toward the east, like Isidore de Séville did, or toward the south, at the initiative of Arab cartographers. In 1500, on the occasion of the great Roman jubilee, Erhard Etzlaub, a Nuremberg citizen, engraved on a wooden tablet the "Romweg" map whose center was the Eternal City, with the two ends being in Viborg, in Denmark, and Naples. Naples occupied the top of the tablet and Viborg the bottom. What seems strange to a contemporary observer is very logical to the eyes of the cartographer: he had envisaged the itinerary that took place through the fixed eyes of a German pilgrim. The representation of the world by Fra Mauro, a Venetian monk and cartographer, was put together between 1457 and 1459, and it too was oriented toward the south. Just prior to Christopher Columbus's transatlantic voyage, Fra Mauro had perfected the art and science of world maps. He took care to integrate recent discoveries as well as narratives from travelers, like those of his compatriot Marco Polo. In Europe, he was also the first to include in a map, even if it was in the wrong place, the island of Zimpagu, the Cipangu of Marco Polo, the Japan of medieval fantasy. Conformist to his system of landmarks, Fra Mauro placed the Pacific coast of Cathay, otherwise known as China, on the left side of the world map. Fra Mauro worked in the service of Alphonse V, King of Portugal. He was influenced by Arab cartography and had mingled with delegates of the Coptic Church of Abyssinia, to whom he owed a portion of the shape of Western Africa. The constitution of knowledge in the Middle Ages was decidedly syncretic.

The determination of an *omphalos* is the particular reflex of any culture that appropriates a territory that is ideally stable and placed at the summit of the hierarchy. Let's go toward the Far West. According to the etymology, *Nippon* (or *Nihon*) means "root of the sun." As a result, Japan became the "Empire of the Rising Sun." This phrase is not just a simple landscape or existential metaphor that insists on the esthetic virtues of a quasi-Homeric dawn with pink fingers. Japan is supposed to be born again every morning at the same time as the sun. In *Écoumène. Introduction à l'étude des milieux humains* (2000), the geographer Augustin Berque examines the case of Japan and of China, both of which, indeed, transport us out of the ecumene by antonomasia, the Greek one. Berque cites Kikkawa Koretaru, a Shinto philosopher from the seventeenth century, who affirmed, "As our country is located to the East, it is obvious that the universal Principle is enlightened before any other country."[27] A priori, there is no doubt about his reasoning. On a map with a 1645 inlay, already influenced by European cartographers and navigators, *Bankoku-sozu* ("map of a Thousand countries"), has an eastern orientation, and Japan is located at the center of the represented space—although the word *Nippon*, which is of Chinese origin, brings a hint of doubt. Japan is perhaps an empire whose axis is the sun, which rises in the east, but it is also a country that is located to the east of China, to which it has answered since the term *Nippon* appeared. The indigenous term for Japan was *Yamato*, which carries the same signification as *Nippon* and which, for its part, does not come from Chinese roots. Whether it wanted to be or not, Japan was integrated into a universe that, as Berque reminds us, was

Sino-centered. China was, and has remained, actually, *Zhongguo*, which literally means "The Middle Country." I could give as many examples as I'd like, but I'll only mention one more here, that of Mesoamerican maps, which adopt the same principle of orientation as the Japanese maps. The rising sun, and therefore the east, is located at the top of these maps. But in pre-Columbian America, the idea of *omphalos* was much more labile, even if the center corresponded to the headquarters of a strong political power. The capital is symbolically represented in the center; it is surrounded by stable natural landmarks, oftentimes hills for the Aztecs and rivers for the Mayas, who inhabit zones with little topographical relief. This type of representation takes into consideration the possible or probable displacements of the population that are to come. The capital occupies a provisionary center, and its privilege is temporary. However, what serves as a border is destined to last, whatever may be the hazards of human adventure that play within the interior of the circle formed by the hills and rivers. If the map contributes to enlarging a political hierarchy—otherwise said the fantasized domination of one group over another in the vein of propaganda—the map also indicates the immutability of the natural environment.

The Composition of Place

Augustin Berque explains that the Sinogram reproducing the toponym *Zhongguo* "combines the figure of an enclosure with, in the interior, those of a halberd (that is to say, warriors), a mouth (a population), and a horizontal line which serves as the earth's surface."[28] The world vision, which determines the Chinese point of view, is not circular, like in Western culture, where the oldest map, conceived of in Mesopotamia, already located Babylon in the center of a cosmos taking the form of a circle. Rather, it is square or rectangular. These squares and rectangles themselves locate their center in the capital of the empire, which then initiates a controlled germination of space and its referents by successive overlapping. The largest squares and rectangles are those that distance themselves the most from the *omphalos*. The largest of all is the one that confines absolute otherness with barbarism. The diagram englobes, rather than opening up, and it limits the space, from where, according to Berque, comes the "inclination of Asia for miniatures of all kinds."[29] The quintessence of this procedure, according to Berque, is the bonsai, in Japan. From an economic perspective, the cultivation of the rice patty participates in an analogous logic. Instead of spreading out the surface of the rice patties by clearing the land, its production is augmented: it is the *nai naru hatten*, "development from within," as stated by the agronomist Iinuma Jirô.[30] In the West, the desire to lay out the space leads to a centrifugal opening onto the world, which is most often translated by the temptation of unjust appropriation. In the Far East, it is a centripetal dispositive of arranging familiar places that takes precedence.

Today, cartographic representation of space revendicates a conformity to the "real," while the medium has progressively dematerialized. This pretension seems to stem from our capacity to reduce the medium to a pure icon, which is practically abstract. The distribution between material and immaterial adopted in the Middle Ages has a quite different twist. The materiality resided in the palpable character of world maps and other media for the representation of space. This representation was not intended to be "realistic." To turn back to this epithet would be unanachronistic. In *The Spacious World* (2004),[31] Ricardo Padrón engages in a very detailed analysis of the manner in which Spain, at the time of its American conquests, conceived of space. Privileging a lexical reading, Padrón finds that, in the eighteenth century, maps and abstract space maintained a large semantic proximity. In the *Diccionario de autoridades* (1726–39), *espacio* and *mapa* were associated. A little more than a century before, Sebastián de Covarrubias Horozco, author of *Tesoro de la lengua castellana o española* (1611), saw the *mapa* as a canvas (*lienzo*) on which the surface of the earth was projected. However, Covarrubias neglected all references to an abstract *espacio*. The Toledan lexicographer, for that matter, was the first to associate map and cloth. As such, the "map" did not exist in the Middle Ages. The use of the term in a geographic sense was not established in French until the sixteenth century. Before, one spoke of *carta*, but just as well *figura* or *pictura*. The earth was represented, but it was not anchored in a stable "reality." The representation of the world was still largely just bits and pieces. The artifact was established; no one thought it was a good idea to dissimulate it. As noted by Ute Shneider, "it is not the geographic position of a place that decides its location on the map, but rather the signification that it has in the context of universal and Biblical history."[32] Real geography is erased, here for the benefit of a symbolic geography, in which the hierarchical order of the time depended on religious references. The world was adapted to the plot that the Christian history conveyed. It would not be a question of realism in the modern sense of the word. It is moreover the real that is established at the discretion of the imperatives of an unchanging representation, that which the Bible and its diverse interpretations have obviously imposed on medieval man. Paul Zumthor, who undertook a brilliant study of medieval space in *La Mesure du monde* (1993), explains this posture: "The cartography of the high Middle Ages proceeded by deduction; taking off from one principle, it extracted elements of a representation from it. Therefore, it explains, it interprets; its objective is to confirm, not to create knowledge."[33] And so it is the biblical space that is confirmed, that of Genesis, of Exodus, that space that Christ himself plotted out.

Egeria was a nun who "came from the extremities of the earth," from the West.[34] She didn't have anything to do with the nymph of the same name who inspired, as the story goes, Numa Pompilius, King of Rome. During her time period, the fourth century, the Western confines of the world were located either in Great Britain or, more surely, in Galicia, which was the archetypal *fini terra*. Around 380, Egeria undertook a three-year pilgrimage in the Holy land, nearly a half millennium prior to the discovery of Saint Jacques's tomb in Galicia and before the Compostela path started to attract pilgrims from East to West. Egeria wrote a summary of her trip and addressed it to "her sisters." But this work was more than just a simple breviary. The *Itinerarium Egeriae*, and especially its first part concerned with the *Peregrinatio ad loca sancta*, would have a great influence on the entire Middle Ages. Pilgrims continued to cite it as an example even in the fifteenth century. Egeria had, materially speaking, roamed over considerable distances. She had visited Jerusalem, but she had also gone to Egypt and Mesopotamia. She had stayed in Constantinople, where she wrote. She did not fail to cross the sands and the mountains of Sinai. It is, rather, the memories of Sinai that start out her narrative. Whereas a modern traveler (such as myself) would unsurprisingly comment on the sands and the mountains, or maybe even the people, Egeria evoked something else:

> This valley is immense; it spreads out from the bottom of the slopes of the mountain of God and it is, based on what we can estimate simply by looking with our eyes and what they say, sixteen thousand feet in length; in height, they give the valley four thousand feet . . . It is the immense and perfectly flat valley where the sons of Israel stayed during the forty days during which Moses went up the mountain of God and stayed there forty days and forty nights. It is the valley where the calf was made, and the spot is still shown today—even a great stone is planted in this spot. Finally, it's the valley at the extremity of which is located the place where Saint Moses led the herds of his father-in-law to graze, where God spoke to him several times as the burning bush.[35]

In itself, Mount Sinai, like all other places that she had encountered, hardly interested Egeria. In its geographic materiality, which she had tested, Sinai continued to pass as an immaterial center symbolized, yet again, by a stone. What really mattered was the spirit that inhabited the place, a holy spirit to which the book of Exodus testifies, which served as a guide to the nun, just as a Baedeker would have done for an English traveler of the nineteenth century. As D. K. Smith notes, having performed a brief commentary of the text, in a more general study devoted to the impact of cartography on English literature at the end of the sixteenth and seventeenth centuries, "for Egeria, as for other pilgrims, the purpose of pilgrimage is not to experience a foreign land in relation to its physical place in the world. In its pure sense, the pilgrimage is taken in order to see the land as a symbol of the events of the past, to read, in some location or physical formation, the historical and religious meaning encoded there. Even the actual places, in these terms, exist purely as symbols."[36]

This perception of space is not a prerogative of the High Middle Ages. It has remained in place for a long time. Without doubt, it lingers still. The vision that Greece has today is yet more closely linked to the stereotype of a solar, insular, obstinately brilliant antiquity that travel agencies and sometimes movie theaters relay. *Mediterraneo* (1991), a film in which the action takes place in the small island of Kastellorizo, in the Dodecanese, enabled Gabriele Salvatores, a Milanese (by adoption) director to obtain the Oscar for best foreign film in 1992. The film follows the evolution of eight Italian soldiers abandoned on their own in 1943. As they are constant reminders of the horrors of war on their Greek island, it doesn't take them long to understand that they would be better off forgotten. Working from the principle that it is better to make love than war, the orderly Farina succumbs to the charms of an innocent prostitute, whereas the sergeant Lorusso organizes a soccer match, while the lieutenant Montini restores an old fresco, and the brothers Munaron, placed as guards at the top of a cliff, return to the joys of a pastoral existence. These images show Greece in all its possible states, similar to what can be seen in Luc Besson's *Grand Bleu*, which was filmed three years earlier. The Hollywood jury was undoubtedly convinced by the perfect balance between the fable of the dominant stereotype of a blue and white Greece surrounded by the waves of the Mediterranean and the charm of the myth. *Mediterraneo* is certainly not an isolated case. The beginning of a list would include *Zorba the Greek*, the Hollywood transposition of *Alexis Zorba*, the novel by Nikos Kazantzakis, as well as *Capitaine Corelli*, taken from *Captain Corelli's Mandolin*, a novel by the British writer Louis de Bernières. We often voluntarily forget that Greece is incorporated into a Balkan ensemble whose geological and climatic specificities are familiar to viewers of the film of Theos Angelopoulos, the greatest Greek film director, one of those who dared to break through the Greek mountains of Epirus or of Macedonia. We can still imagine the travelers who came in *Magna Grecia* in the eighteenth century. Southern Italy did not interest them in the least, if we believe the narratives of their voyages. What really caught their attention was being *there* while all the while being *somewhere else* or at *another time*: temples, vestiges, and mythological sites. Alain Corbin speaks in relation to this of a "subordination of the regard of the Latin text," which "contributes to the explanation of the absence of an authentic descriptive style."[37] And to evoke the voyage of Addison in Campania, who took as guides for the trip out Horace and Virgil, and for the return trip, "'why try to paint what they have done so well?' Addison asks, who offloads on Horace, Virgil, Stace and Lucain the care of describing the landscape."[38]

The English traveler, a great journalist ahead of his time, had traveled around in the Mezzogiorno during the very first years of the eighteenth century. But in his *Voyage en Sicile* of 1788, Vivant Denon proved that the classical approach had hardly evolved. He rarely mentions Sicily, which was contemporary to him; he goes just as far as to transcribe the names of the places in the Greek manner. He was happy to substitute the Latin text for the Greek one. Goethe might have been the first to *see* the reality of Mezzogiorno in his *Voyage en Italie*, completed between 1786 and 1788 and published nearly thirty years later.

The Orient of the pilgrims and the crusades takes on a particular color. It is both an authentic exotic space and a space of the Scriptures, which is totally familiar. That brings us back to what has already been said: the center of the Christian world, when it is about Jerusalem and not about Rome, is indicated by an *omphalos* that is situated outside of the familiar space. This converges to augment the heterogeneity of the environment while narrowing the vision of the space around an infinity of local households, more or less independent of each other. From the tenth century, and especially the first half of the twelfth century, Saint-Jacques of Compostela offered a real alternative to Rome and to Jerusalem. The supposed tomb of Saint Jacques had its turn at being a type of *omphalos*, and the *Christianitas* suddenly found themselves with two axes, which, by a remarkable paradox, implicated both the Oriental and Occidental extremities. Most of the time, the pilgrims pursued their voyage beyond the *Campus Stellae*, which gave its name to Compostela, to go to Finisterre. There they bathed in the cold, agitated waters of the actual Costa de la Muerte, in an effort to purify their bodies. The environment escaped from a truly global vision. It was difficult to imagine the world. Impossible to know where it ends, or how: its geographic form has to be guessed. With the exception of scholars, only a marginal attention is paid to its physical configuration. If it is perceived as a whole, this universe is that one of the Christian God, who was not a geographer by vocation. The space is made up of an ensemble of local entities, whose limit is fixed by a homogenous perception of communal experience. Mary B. Campbell notes, rightly so, that in *Itinerarium*, for Egeria, there are not "places," but "places where"[39]—that is to say places where something has taken place for the pilgrim from far away and where something has taken place for those who live a more limited existence. Place materializes itself in the prolongation of a past event and in the collective present. As far as concerns its isolation, place is in harmony with the subjective harmony of those who live it, of those who live at it. According to D. K. Smith, for whom certain maps privilege "the relationship to ideas rather than to space,"[40] this encompasses an especially emblematic value because there is not an articulation between the different sites. Smith gives the example of the romance of Sir Bevis of Hampton (Beuves de Hantone). Bevis is the hero of an Anglo-Normand epic from the thirteenth century, adapted many times between the thirteenth and fifteenth centuries and even into Yiddish at the beginning of the sixteenth century (the *Bovo-Bukh* is the oldest nonreligious classic in Yiddish literature). Bevis of Hampton lost his father, the count of Hampton, who was assassinated in a forest by assassins hired by his own wife and the Emperor of Almain (Germany). Just like Hamlet, the young Bevis swore vengeance. A little after the incident, which was followed by the marriage of the countess with the emperor, Bevis just barely escaped a new plot against him set up by his mother. He survived but was nevertheless sold to pirates who took him to a South whose geography was variable: Armorica in some versions, Egypt in others. After multiple combats and sentimental adventures, which took him all the way to Damas, Bevis returned home and carried out his revenge before being condemned to exile. He saw many countries and was able to describe the world of his era, but he didn't do anything. For the author or authors of the text, places are names rather than concrete references. As D. K. Smith states, "mention is made of Almaine, Damascus, Babylon, Jerusalem, but while the names evoke a sense of unspecified and exotic distance, there is no attempt at placing them in the world."[41] These toponyms are the markers of foreignness, the referents of which are hardly more real in the minds of contemporaries than the monsters that haunt the distant margins. Medieval texts are all submitted to an analogue rule. It is the same for Constantinople and Jerusalem in *Voyage de Charlemagne*, an epic from the twelfth century, or the island in *Nibelungen*, from the thirteenth century, for example.

In one of the stimulating studies collected in *Mimésis* (1946), Erich Auerbach gave quite a bit of attention to the *Chanson de Roland*, whose Anglo-Normand manuscript, labeled "Oxford," is considered to be the original and dates back to the end of the eleventh century. According to the German philologist, the structure of the poem, conforming to the representation that one held of the world at that time, was made up of a sequence of independent scenes: "The need to establish links and to pursue a development is weak; even from the interior of scenes, the development, when we find one, is laborious and full of hesitation; but the gests of the scenic moment are endowed with a strong plasticity which can only impress the reader or the listener."[42] It is the gests and the attitudes that are put into relief; the sequencing of the moments is secondary. Like a sentence, time comes from the parataxis: it is lacking coordination, coordinators. The link is not even conceivable. The present throws back to sacred history that confers on time a superior, supreme, figural homogeneity, within Auerbach's vocabulary. It goes the same for places. Each one among them functions like an independent cell, because, anyway, according to Auerbach, "the fall of the Roman Empire of the West and the order which it incarnated signifies also the disintegration of the *orbis terrarium*, in such a way that a new world could only rebuild itself from the fragments which are disassociated from the previous unit."[43] Places are the fragments of a lost whole, and there again the sacred texts convoke a geography of the whole that is no longer recognizable and feeds the nostalgia of origins. Medieval spatiality participates therefore in a paratactic dynamic, in the manner of narration and of temporality. The passage to a spatial approach, which will go as far as to integrate the links, to go beyond the parataxis for the benefit of a concatenate vision of places, will come later. Hypostasis constitutes one of the most eloquent signs of the passage to the Renaissance.

In this universe devoid of clear coordination, the relation to the referent is so labile that the distinction between a real voyage and an imaginary voyage is sometimes erased. For those who cannot undertake a long trip, substitutes, which are completely acceptable, are conceived, such as the maps of spiritual pilgrimages. Even precise examples are taken into consideration. In the year 1500, Johann Geiler von Kaysersberg, a Swiss humanist that D. K. Smith was quite right in exhuming,[44] imagines the situation of a prisoner who cannot make the pilgrimage to Rome. Does incarceration justify an exemption? Certainly not! Geiler calculated that on the outbound trip, a pilgrim would take 22 days to get there, to which is added 7 days to visit churches. The return trip would be spread out over 21 days. Altogether, it would take 50 days to make the trip before returning home. So the prisoner would know what to do to obtain spiritual compensation: he would need to walk 7 miles a day in his cell during 50 days.

Even those trips whose objectives are other than religious are to be considered with caution. The most popular travel narrative of the Middle Ages is uncontestably the *Livre des Merveilles* (1357–71) in which Jean de Mandeville, who was undoubtedly a native of Liege but who was also known under the name of Sir John Mandeville since he called himself English, told the stories of the trips he made starting in 1322, across Egypt, Russia, China, the Indian Ocean, and many other places, some of which were even more prestigious and fabulous than others. But did Mandeville even exist? Nothing could be less sure. It's always the case that the more or less "realistic" relationships are associated with the considerations of Noah's Ark and the Kingdom of Prester John, in a logic that is comparable to that of Egeria. Chapters 21 and 22 of the *Livre des Merveilles*, which speak about Java, its king, and its inhabitants, are particularly successful, from the point of view of amateurs of *mirabilia*. For example, we learn that the island of Dondun is located to the south of Java and its king reigns over 54 islands and just as many kinglets, whose subjects present diverse deformities. The first among them is populated by cyclops. That leaves 53 to describe. It is futile to insist: we can imagine the *crescendo*! To get back on track with a more acceptable geography, it would be useful, after such a reading, to undertake the geography of Ibn Battūta's work *Voyages*. Leaving behind his native Tangers, Ibn Battūta dedicated the entirety of the second quarter of the fourteenth century to discovering the world, as no one before him had done: he went to Russia, India, Sumatra, the coasts of what is today known as Kenya, to the Mali Empire, and so on. Whereas the existence and the journey of Mandeville are contested, Ibn Battūta *really* travelled. For the record, we can note that it was Mandeville who influenced Columbus, and not Ibn Battūta, the great Tangerian. If it had been the contrary, Columbus might have discovered the Indians, and the Caribbean "Indians" would have perhaps survived.

Babylon and Taprobane at the Intersection of Worlds

In the Middle Ages, people set their eyes on an East that was more or less close . . . or far. The pivots of an indecisive geography, which serves as a foundation for the Christian world, are located in Rome, but also in Jerusalem and in Constantinople. Two of the three mainstays of the *Christianitas* are on the edge of a wonderful universe whose extension challenges the speakable. Many guides confirm the incursions of the medieval man in the vast spaces that are now forgotten. Among them are the prophets of the Bible and the great figures of Greek, or even more so, Latin knowledge. The hierarchy of places that these authorities competed to establish is unstable. It evolves in function of the place of each one and the others in the canon of the moment. The projection in the real or imaginary space is a good tributary way to measure the reception that is accorded to the texts and their authors. Time establishes the classics, and without doubt it is time that designated masterpieces. In any case, time activates the trends and enables openings. As for the cleavage between the real and the imaginary, it covers only a limited reach. The trip is often too mental for the reference to reality to be objective and for the places to be dependable. It is the text that ensures the link between the prestigious past and the irresolute present. Certain sites have lived on in memories because their name was consigned in the pages of the Scriptures. Babylon is such an example. This Old Testament structure is known. Other places start to show up in the pages of treaties as prestigious as Pliny the Elder's *Histoire naturelle*, whose reputation has never been forgotten. These places were random, because the Middle Ages had lost their coordinates. These are toponyms exempt of referents, cities or lands deprived of mooring, which float somewhere on the margins of the conceivable. These are the markers of infinite space. One of the most significant is Taprobane, the beautiful name of which has traversed centuries and cultures, intriguing a quantity of scholars.

Before the thirteenth century, journeys to Mesopotamia were rare and Babylon remained a figment of literary imagination. As it so happens, it was not the Christians who undertook these voyages but the members of the Spanish Jewish community (Benjamin de Tudèle) or of Bavaria (Petahiah de Ratisbonne) and adventurous Arab erudites such as Ibn Jubayr, whose surname was al-Balansī, the Valencian. They arrived in Babylon by crossing the sands of Iraq. In the Middle Ages, Babylon had become a pile of rubble. Not surprising, then, that the stories are sparse. But there are still a few. Ibn Jubayr went to Iraq. In the spring of 1184, he made a stopover at Koufa and at Hillah, in the immediate surroundings of Babylon, which he abstained from mentioning in his work *Rihla* (his *Voyage*). Just before him, Benjamin de Tudèle had been to Babylon to take the census of the Jewish communities and the synagogues. The narrative of these travels, written in Hebrew (*Séfer-Masa'ot*), published in Constantinople in 1543, were translated into French by Jean-Philippe Baratier and appeared in Amsterdam in 1734.[45] In reading the rabbi of Tudèle, we can understand why Ibn Jubayr eluded the places. They were deprived of interest the moment that one looked for something other than biblical vestiges. Benjamin notes the tower that is located four miles from the city. He briefly recalls the existence of the city of Hela, at a distance of five leagues, where he counts ten thousand Jews and four synagogues. For Benjamin, as for Ibn Jubayr, the present time city of Hela (or Hillay) was tangible in other ways, different from the souvenir of Babel. As for Babylon, it had become a large empty circle, whose circumference it became a game to guess, a *locus horribilis*, infested with snakes. Babylon would not have more success with Ibn Battūta. At the end of the 1320s, the Tangerian had probably reached the spot. Like Benjamin de Tudèle, he had few words to say about Hillah, which had passed into the hands of the Mongols in 1260. And the words he used belonged to Ibn Jubayr. Although he was a great traveler, Ibn Battūta did not hesitate to plagiarize his predecessor.[46] A philosopher, he went as far as to say, "But God knows best the truth in all of this."[47] And God also knows what must be given to Babylon: much less than envisaged, if the Arab travelers are to be believed. We would have to wait until the middle of the fourteenth century for the Christians to venture out to these areas. But their objective had nothing to do with that of their Jewish and Muslim predecessors.

At the same epoch, Taprobane stopped being so attractive to the Christian imagination, even if it is mentioned in Isadore de Séville's *Étymologies*. The ring of this name held meaning for only a few aware readers. In days gone by, Taprobane had, however, found its way into the antique Almanac through the intervention of Onésicrite d'Astypalée, who, upon the return of Alexander the Great's campaign in India, testified to the existence of an island off the southern shores of India where elephants were numerous and savage. This island could be today's Sri Lanka. Onésicrite was both an officer and a historian. His *Education d'Alexandre*, written around 320, is lost to us, but it inspired commentaries on the parts of Strabon and Pliny the Elder. And these works are available to us. A little before the Christian era, in book II of his monumental *Bibliothèque historique*,

Diodore of Sicily recorded the surprising adventure of the merchant Iambulos. Together with another Greek man, Iambulos was successively captured by Arabs and then Ethiopians. The latter gave the two men the order to sail in a direction of a distant island. The purification of the lands of their abductors would depend on the success of their mission. They were provided with a boat and supplies. At the end of a four-month crossing, Iambulos and his companion arrived at Taprobane. The hardships would just as soon be over, because the island was fertile and its clean-shaven inhabitants were very hospitable. The two Greeks from Diodore stayed seven years on the island. The *Odyssey* clearly haunted their minds. This trip was pursued through many unknown works. Traces of Taprobane or of its avatars can be found in Eratosthenes, Pomponius Mela, Arrien, Solin, and others. But let's go back to the Middle Ages.

From an Occidental perspective, it didn't take much for the year 1244 to make its mark on history as much as 1492 did. Hinting at a shift in geopolitics, this year nearly engaged a first movement of globalization. In 1245, the Lyon synod placed on its agenda the question of the Tartar invasion in Oriental Europe. Pope Innocent IV sent a series of legates abroad in an effort to make the Mongols their allies against the Arabs, who had just reconquered Jerusalem, and, while there, to convert them. It was an ambitious program, destined to fail. While the Mongols had a sense of hospitality, they were also fast on the draw: they, for their part, asked the pope and the kings and emperors of Europe to submit. In 1245, Jean de Plan Carpin left for Karakorum, the Mongolian capital located to the southwest of Lake Baikal, before Kublai had it transferred to Dad (Peking). At the initiative of the pope, Jean de Plan Carpin was followed shortly behind by Laurent de Portugal, Ascelin de Lombardie and André de Longjumeau. These legates were imitated in 1254 by Guillaume de Rubroek, who affirmed in his *Itinerarium ad partes orientales* that he had crossed the border of China. It is pointless to insist on the exploring voyages of the Polo family. Nicolò and Maffeo arrived for the first time in China in 1266 before returning there again in the company of Marco, the son of Nicolò, between 1271 and 1298. As is often the case with a book that merits the glory of a traveler, *Le Devisement du monde* came from the voyage that the Venitian made with Rustichello of Pisa, with whom he spent several months in captivity in Geneo.[48] That is the historical context in which Polo evoked Java Minor and its eight kingdoms in chapter 167 of *Le Devisement du monde*. He contributed, certainly, to bringing Taprobane back to the surface, but on the other hand, the location of the island became more complicated. Without ever making a reference to the Greek toponym, Polo populated the most eastern part of the Oceanus Sea with islands. Taprobane was no longer an isolated island on the outreaches. Today, it is considered that Java Minor was Sumatra. This island can be distinguished from Seilan, another island that Polo was the first to indicate under Arab consonants. Even though its surface area has been reduced by the rising water levels, Seilan was "truly the best island there was in the world,"[49] according to Polo in chapter 174. The enchantment of the place comes first from the infinite variety of gems, which they held—"because let me tell you on this island, noble and great rubies are born, and in no other part of the world are they so splendidly developed. And there are sapphires, and topazes, and amethysts, and garnets, and many more other stones."[50] The king possesses the most beautiful rubies in the world, which Marco Polo could see with his own eyes. Among the emergence of new islands, Taprobane was prone to shifting back and forth between Sumatra and Ceylon, whose dimensions corresponded more or less to those of the antique model. The island of Onésicrite had reintegrated the imagination of explorers.

In the Arab world, the island was well known. Legend holds that the fall of the first man took place at the summit of Sérendib, called since then "Adam's Foot." This mountain would give one of its names to the island, where Ibn Battūta set out in 1344. The traveler had a discussion with a local "sultan" in the Persian language and undertook a pilgrimage at Adam's Foot. His narrative is lively and detailed.[51] Starting in the 1320s, a couple of Franciscans roamed across South India and inaugurated a diocese. One of them, Jourdain of Sévérac, cites Ceylon in his *Mirabilia descripta*. In 1348, another Franciscan, Giovanni Marignolli, wound up there after a storm. He stayed in Ceylon in a type of captivity for nearly four months and also went himself to Adam's Foot. His writings were only just discovered in Prague in the eighteenth century. They had no influence on his contemporaries. Lacking a specific knowledge of the terrain, stories were invented, even political apologues. Good stories can be found, for example, in chapter 33 of the *Livre des Merveilles* by Mandeville, who located Taprobane as an island next to the kingdom of Prester John. The island was open to Christians and overflowing with riches. There were two harvests a year. There were mounds of gold, which giant ants watched over, but there were ways to distract their attention. In *Les Remèdes aux deux fortunes* (1360–66), an allegorical dialogue contemporary to Mandeville's narrative, Petrarch made Taprobane into the headquarters of an ideal democracy where the king was elected on the basis of his virtue and not according to blood laws. Such a marvel could only occur in this corner of the world where, two-and-a-half centuries later, Tommaso Campanella set up his *Cité du Soleil* (1602). The influence of Marco Polo would not, undoubtedly, suffice in establishing Taprobane within a self-proclaiming credible geography. But a considerable event took place at the turn of the fifteenth century. Invited to teach at the University of Florence, Manuel Chrysoloras arrived in Constantinople in 1397. He brought along in his large suitcases the *Geography* of Ptolemy, a Latin translation of which he would publish in 1406. Thirteen centuries after it was written, the

Christian West finally discovered the text equipped with a set of maps that would revolutionize its vision of the world. Ptolemy mentioned Taprobane and gave its measurements. Therefore it exists. But doubts persisted: was it Ceylon or Sumatra? It was most likely Sumatra, to tell the truth.[52] Cartographers have confirmed the hypothesis.

Taprobane finally emerged from the mist of warmth that had enveloped it for ages. One day, it would even come out of a cloud of dust from the English Channel and still yet the imagination of Don Quixote. In chapter 18 of the first volume of his adventures, Ingenious Hidalgo believed he was seeing off in the distance the army of the furious Prince Alifanfaron of Trapobane (the rhotacism is his own) waging war on the king of Garamantes. As for Babylon, it sunk further into the sands of the dessert. And while Taprobane was being pulled between Ceylon and Sumatra, the city in ruins came off its hinges. This brought to mind the existence of an alternative Babylon in lower Egypt, on the Western Bank of the Nile, a little to the north of Memphis. This Babylon was engulfed by Cairo, which was developed between the ninth and the end of the twelfth century. That's where we were when Mandeville undertook his problematic trip. What did this man from Liege see? Egypt and Palestine, maybe; Taprobane, certainly not. As for the rest . . . To fill in the intervals, he opted for a radical solution. Instead of plagiarizing his precursors who were deemed credible, he made everything up by doing his best to prove his objectivity and worth, this paradox being less shocking in the Middle Ages than it is today. In chapter 6 of the *Livre des Merveilles*, he says that he stopped at Babylon. As planned, his Babylon was the Egyptian one. It borrowed many traits, however, from its big Mesopotamian sister. It also held sacred vestiges. Of course, Mandeville distinguished the two Babylons: the one of the banks of the Euphrates continued to swarm with snakes. To recap, the stay in the Egyptian Babylon enabled Mandeville to evoke the other Babylon, which he had neither visited nor even gotten near to, in order to establish a list of pertinent stereotypes: the presence of serpents and the immense space that was only a vestige. After Mandeville, the rhythm of real or imaginary stays in Babylon diminished.

En route for India, in the second quarter of the fifteenth century, Nicolò de'Conti stopped in Babylon. For him, the city was called Baldacco. But it so happens that Baldacco was the Italianized name of Bagdad, and the etymon of a four-poster bed. "On the heights of the city there is a fortress and a royal palace which is beautiful and solid. The king of this province is very strong,"[53] he adds, in confusing the two Babylons and Mandeville. From the double Babylon of this former, we arrive at a Babylon that subsumes a part of the East. Then it was no longer spoken about. But it will be spoken about again starting from the end of the sixteenth century. Among those who exhumed the city from sand and from oblivion, there was John Eldred, in 1583. Was this Eldred unknown? It's not so sure: one of the witches from *Macbeth* makes a reference to his trip (1.3). In the meantime, Taprobane affirmed itself. Nicolò de'Conti certainly went there rather than to Babylon. He gave, in any case, a precise description of cinnamon and mentions pepper, camphor, and gold. But the inhabitants worried him. Men and women both had large ears from which hung precious stones. Some were cannibals and headhunters. Wealth was measured by the number of heads one had on display at home. Taprobane was, for him, Sumatra. After Vasco de Gama had, thanks to the indications of Ahmed ibn Mājid, his Arab pilot, reached India by waterway, the island hosted a series of exotic tales. In his *Itinéraire* (1510), Ludovico de Verthema was the first Westerner to enter into Mecca after his conversion to Islam, and to speak about it, like Maximilien of Transylvania who told the story of Magellan's trip to the Cardinal of Salzbourg.

For all these men, Taprobane continued to be Sumatra. Even if Ceylon was more agreeable, only Andrea Corsali, who discovered New Guinea, preferred to see Taprobane, "where topazes, hyacinths, rubies, sapphires, carbuncle, agates, garnets and beryl are still present."[54] This striking list dating from 1518 seems to have inspired the ancient Greeks and Polo. But the biblical intertext is in filigree. In chapter 28 of Ezekiel, God commanded the prophet to announce his degeneration to the king of Tyr who was thrilled at the idea of the fall of Jerusalem. Announcing to the sovereign his destitution of the next day, Ezekiel exclaimed, "All sorts of precious stones will make your coat: sard, topazes, diamonds, beryl, onyx, jasper, sapphires, emeralds."[55] The ground of Taprobane was like the coat of the king of Tyr. Corsari knew the Bible as well as the *Le Devisement du monde*. As for the tomb of Ezekiel, it was located on the Euphrates, not far from Babylon, as Benjamin de Tudèle remembered. Still at the beginning of the sixteenth century, in the big blur of spaces filled with text, Taprobane is not so far as Babylon.

Who, in the Middle Ages, had best understood the nature of these spaces? Ptolemy, to whom is attributed this posthumous homage? Cold. Ibn Battūta? Warm. Marco Polo? Hot. Kublai? Burning hot. Did we have to wait until the publication of *Les villes invisibles* (1972) by Italo Calvino to understand it? When the postmodern Marco Polo tells the story of the empire to Kublai, his catalogue of cities is as poetic as that of the geographers of long ago. But the emperor perceived that all the places resemble each other. So his mind starts to wonder "and, once the city is taken apart piece by piece, he rebuilt it in another way, by substitutions, displacements, inversions of their ingredients."[56] Only Polo is left to verify if the cities invented by Kublai exist: "Come back and tell me if my dream corresponds to reality."[57] In Babylon, which Grand Khan knows is threatened by incubus and maledictions,[58] and in Taprobane, more than one traveler has tried it. And each of the audacious were able to conclude, like Kublai that, "maybe . . . the empire . . . is nothing other than a zodiac of fantasies of the mind."⁵⁹ What had Babylon and Taprobane been in the Middle Ages? It is difficult to say because the perception of the place varied according to points of view. At this time when the spatial approach of the *Christianitas* was paratactic and places were conceived of outside of all correlation, Jewish travelers (in Babylon) and Muslims (in Babylon and in Taprobane) seemed better equipped to face the fantasies of the mind. The identification of sources, especially when confessional, is capital. The representation of places depends on focus. Another observation emerges. The place is constructed according to an intertextual logic, because the narrative of one predecessor, if they have an established reputation, creates a credible base for the development of description. Antique and medieval geographies work in this way.

The Philology of the New

The randomness, and for the contemporaries the quirkiness, of the coupling between "real" geography and imagined geography endures in diverse forms. Was it necessary to invent a place? Was it necessary to represent Babylon and Taprobane? Why not make the place? In 1540 Ignace de Loyola, a hidalgo having become a saint, founded the Company of Jesus, the first foundation of which, *Spiritual Exercises*, didn't take long to be hallowed by the Vatican. Rule 47, which is part of the first spiritual exercise, establishes that before undertaking his meditation, the scrupulous Jesuit should deliver himself to a *compositio loci*. This is how Igance describes the process:

> The first prelude is a certain way of organizing the space [*compositio loci*]. To this effect, we should note that, in any and all contemplation or meditation on a corporal reality, Christ for example, we must represent for ourselves, according to a certain imaginary vision, a corporal place representing that which we are contemplating, like a temple or a mountain, a place in which we can find Jesus Christ or the Virgin Mary, and all of which is concerned by the theme of our contemplation. If, on the contrary, a non-corporal reality is scrutinized, as is the consideration of the sins now proposed, the construction of a place could be to imagine that we see our soul enclosed in this corruptible body as if in a prison, and the man himself exiled in this valley of misery in the middle of animals with no reason.⁶⁰

This precept merits a rather long commentary, because, indirectly, it introduces some of the major ideas linked to the treatment of space in a representation system that couples the real and fiction. It almost puts forward the topoanalysis posited by Gaston Bachelard, "the systematic psychological study of location of our intimate life."⁶¹ For Ignace, it would seem that "objective" space no longer exists. It lies entirely within the mental composition one makes; actually, it is in the representation that one makes, that one makes for *one's self*. Therefore, the distinction between a "visible thing" and an "invisible thing" fades. Certainly, the examples included or suggested in rule 47 are disconcerting. Burned by the Romans in 70 AD, the temple of Jerusalem could hardly appear more "visible" than the sins, in the sixteenth century as today. As for the "invisible," it is not limited to hell, toward which rule 65 abounds with delight: "The fifth exercise is a contemplation of hell [. . .] The first prelude contains here the composition of the place, by putting before the eyes of the imagination the length, width, and height of hell."[62] The exercise, consisting in the mental composition of chosen places, emphasizes the iconic virtues of representation, its iconographic potentialities. Indirectly, he points out the relationship that links the representation to a referent, this referent being a stranger to the objective reality. But does hell exist? Never mind! What matters is not so much the thing in itself, the thing as far as it "exists" (and anyway, *how* does it exist?) but the idea, or the composition, that one has made for oneself by direct observation or by an effort of abstract reconfiguration. In a manner not in the least biased, the spiritual exercise suggested by Ignace attempted to demonstrate that the history of the representation of places is an integral part of the history of ideas. It is true that, in the Western part of the terrestrial orb, as elsewhere, religion has long while ensured the unity of the spatial vision. According to *Scriptus per digito Dei*, if we believe Pier Damiani, an eleventh-century monk from Ravenna, the world was the playground of the soul, the theater of his earthly adventure. We can observe that the term *vision* is appropriate. While Ignace advocated composition, Theresa d'Avila drew sketches of her visionary projections and Jean de la Croix drew Mount Carmel as if it were a type of precursor of a *Carte de Tendre* of the soul.[63] These three Spanish Saints were preceded in their practice of spiritual voyages by Maxime de Tyr, a Greek Neoplatonist, Denys d'Alexandrie, but also Aristéas de Proconnèse in the sixth century, who, according to Christian Jacob, "had the power to let his soul ramble by mountains and valleys and to acquire therefore an unlimited science on places, airs, and waters, the human world just as the world of stars."[64] The old myth of Icarus and Daedalus falling from spaces does not therefore cease to be repeated.

The extreme lability of medieval space, and sometimes Renaissance space, is explained by different reasons, the first of which is its surprising nature: space has not always been a spatial concept. Paul Zumthor has undertaken an enlightening study on the philology of *spatium* in *La Mesure du monde*. In this way he reminds us that medieval languages did not possess a specific term to express the concept of space. A rapid survey of the etymology of *spatial* indicates that the neo-Latin aspect breaks down into two lines of thought: that of *locus* and that of *platea*, in low Latin, one of which evolved into *lieux*[65] and the other into *place*, which the English language then adapted it in its own way. From a Germanic angle, the term *rum*[66] was chosen, which the Swedish language conserved in its actual state but which German transformed into *Raum*. When we consult etymological dictionaries, we find that the Latin *spatium* has not undergone a popular evolution, but it ended up by lending itself to the learned construction of the word *space*. The first occurrence of the word is found in the *Chronique Ascendante des Ducs de Normandie*, in the *Roman de Rou* of Wace, a Jersey author who wrote in Anglo-Normand. Dating the beginning of his narrative in the *incipit*, Wace indicates that "*mil chent et soisante anz* [1160 years] *out de temps et d'espace*" have passed since the birth of Jesus Christ. As we don't know when Wace completed his work, we can pinpoint the origins of this *space* between 1160 and 1174. For Wace, the "space" was more a lapse of time, a duration. It's around 1200 that it took on, for the first time, the meaning of an "extent" in the *Dialogue Grégoire*, but it was only a matter of the *lo spaze del corti*, the "extent of the court." Space was still hesitating between place and time. This indecision would last until the sixteenth century.[67] It was based on what was happening in Latin, where the *spatium* was also a racetrack and, from there, a circuit of an extension of time, an interval. Many Latin words had been invested in the same lexical field. Whereas *spatium* was a racetrack, *spatiari* designated the action of the *spatiator*, an individual who walks far and wide, a *flâneur*. There again, we hesitate, between racing and dawdling, between speed and the praising (or condemning) of slowness. In modern Italian, there is the verb *spaziare*,[68] which conserves this sense, even if it is only in an abstract way: it is thoughts that *spazia*, that "vagabond." In Spanish, "slowly" is said by *despacio*, an adverb that derives from *de espacio*. To make progress, to make headway, is to go from one space to another, to fill in the gaps, slowly. But let's head back to Rome. During the Imperial Era, the verb *exspatiari* meant "to deviate," "to spread one's self far away." The conquest was, in a way, a sort of normalized *exspatiari*.

We would have to wait until the seventeenth century and Descartes's *Principes de la Philosophie* (1644) written in Latin (*Principia philosophiae*), for the abstract meaning of what we consider today as "space" to affirm itself. The term *spatium* is found throughout the entirety of the Cartesian text: I counted about 150 occurrences! In the meantime, in the sixteenth century, a century in which the gaze was traveling further and further, higher and higher, horizontally and vertically peaking, space was also becoming a "stretch of air." Basically, *space* was a spacing, a distance to be traveled. The "space of a moment," as one says in modern French: the word has conserved something of its origin. According to Zumthor, "medieval space is therefore that which is between two: a void to fill. It can only be made to exist by peppering it with sites,"[69] just as we've seen on the maps I've mentioned and in the romance of *Sir Bevis of Hampton*. Medieval spacing takes us simultaneously to a "hole-y" vision of space and a temporality, a "space-time." There is a tight association between the philological itinerary and the process of "reading" of space, which is not surprising. Taking into consideration, in a simultaneous way, space and time (spatial-temporality), which constitutes one of the major founding principles of geocriticism, is not an option in the Middle Ages but an unavoidable reality. Space is inscribed in time, in the diachrony, in history.

Hereford is a small town in Arizona where a group of cowboys, having defied Wyatt Earp and his gang during the famous shoot out at OK Corral, met up. Hereford is also a small town in Texas where Ron Ely was born; Ely, at the end of the 1950s, played a small part in the series *The Life and Legend of Wyatt Earp*, before becoming a famous television Tarzan. Hereford is finally also an English town whose cathedral is decorated with one of the cartographic wonders of the Medieval Era: the world map of Richard of Haldingham, who, deviating from the practice, signed his work. Adopting the T-O model, the *mappa mundi* of Hereford made Jerusalem the *omphalos*, as did the map of Ebstorf, made fifty years prior. Christ reigns at the summit, and Judgment Day is located outside of the borders of the world and therefore outside of space-time.

Richard of Haldingham materialized the spatial-temporal vision of the world that was contemporary to him. His map exclusively accounted for not only the known or imagined geography of the universe but also the history of this geography, which, of course, could be traced back to sacred texts. The stake was high and supposed that the document conveyed a diachronic image of the world, a temporal depth that had to be rendered compatible with the surface planes of the vellum. Long before computing and computers, the notion of hypertext made its appearance in the cartography workshops of the Middle Ages! Richard took recourse in many strategies to expose his vision of things. He started by annotating history in the margins of the reproductions of lands and waters. His world map bore forth an abundant text that described and commented on the symbolic value of the sites it referenced. But it also contributed to giving a temporal depth to the global perception of the world by submitting the spectator to a series of anachronistic juxtapositions, or more precisely, "polychronic" juxtapositions, because they testify to the copresences of temporal references that are little or are unlikely to be compatible. In geocritical vocabulary, we can speak here about "stratigraphic space." On Hereford's world map, the Tower of Babel and Medea's Colchis cohabitate, just like Saint Jacques of Compostella and Mont Saint Michel, the Abbey of which was built beginning in 966. We have noticed that five places govern the structure of the world map: the Garden of Eden, Babylon, Jerusalem, Rome, and the Herculean columns. These five points are quasi-equidistant, and, according to Nathaniel Harris, "they evidently form a kind of implicit narrative, tracing humankind from its beginnings, through key episodes in the old and New Testaments and the triumph of the Church."[70] The itinerary of "this kind of spatial narrative"[71] starts off at the center (Jerusalem) before heading toward the banks of the Oceanic Sea, beyond which reigns the void . . . or a solitary God. In any event, the distinction between the world map and the romance or epic (among other literary genres) is weak: one is a visual narrative, the others are textual maps; neither one nor the other absolves space from time.

The desire to record the spatial trace of history through the intermediary of the map was not unique to proto-European *Christianitas*. In pre-Columbian America, which, on the world scale, constitutes an alternative to consider among others, the Aztecs and the Mixtecs proceeded in a similar hybridization between space and time. For the rest, the choice of the term *hybridization* is not very welcome: the dissociation between temporal and spatial dimensions is a recent phenomenon and is somewhat marginal if we measure it in the light of the very long history of the ensemble of the planet. The spatial representations that the Aztecs reproduced on the *lienzos* of the conquistadores are composed, as a general rule, of three series of indications: those that concern the geographic description of the territory, but also others that indicate something about the mythical origins of the community, and then also the chronological list of chiefs. The latter, as is the case for the world map of Hereford, occupies the margins of the territorial figuration itself. According to Elizabeth Hill Boone, the Aztec map "represents pictorially or conventionally not what is seen but what is understood to be."[72] Consequently, the determination of a location doesn't carry geographic information in and of itself but serves rather to tie the event to what the American anthropologist calls "sites of memory."[73]

The map situates the history of a people in the territory that is assigned to it, or within the territory it has attributed to itself. It is therefore about reconciling the historic diachrony and the spatial synchrony. The Aztec approach is far from being isolated. Elizabeth Hill Boone cites the case of the Western Apaches and the Arandas, Australian Aborigines about which I speak myself in the prestigious wake of Bruce Chatwin. To tell the truth, Elizabeth Hill Boone could have mentioned certain households coming from the very exotic Occidental Europe of the Middle Ages.

CHAPTER 2

The Horizon Line

The Opening Up of Space

The distinction between place and space has been examined from all angles, all latitudes, and all seams, without coming to satisfying conclusions. The mere abundance of all the definitions indicates that there is a persistent doubt about the criteria of demarcation. Rather than going by my own antiphony, I would prefer to make a brief stopover via the *Odyssey*, a most efficient last resort in case of hesitation. Penelope, the sedentary woman from Ithaca, and Ulysses, the nomad of the oceans, incarnate two complementary visions of space-time. The geography traced out by Penelope anticipates the medieval *spatium*. Immobile on her island and immobilized by the suitors who are sieging her palace, the queen is reduced to imagining a subterfuge: she weaves a canvas whose completion corresponds to the hour of her surrender. In fact, Penelope gives herself over to an equation whose parameters have too many unknowns. How much time does Ulysses have to get home (as long as he does) before the suitors definitively lose their patience (as long as they are courageous enough to take action)? It is impossible to answer these questions. Penelope is only sure of one thing, which upsets her: the time required for Ulysses to return home is longer than the time required to weave a canvas, despite how large it is. It is therefore necessary for Penelope to come up with a plan and even "string along tricks" (XIX, 137). She stretches out the space of time between the beginning of the weaving and the completion of the shroud that will one day cover the corpse of Laertes, the father of Ulysses. We know what happens next. She undoes at night what she weaved during the day, as long as the suitors are willing to wait—long enough, she hopes, for Ulysses to return again to the shore of Ithaca and the cobblestone floor of his palace. The queen conceives of the spatial space (we now know that is not a tautology) as a space of time, from an absolutely static position, because she is prisoner in her own home and the only intimacy at her disposal is the surface of her marital bedroom. The range of her projection within geographic space is more or less zero. She tells the beggar (who is Ulysses in disguise) that during three, nearly four, years she settled for reproducing the back-and-forth movements of her husband's sailing on a loom. She went forward during the day and retreated by night. Her loom is a metaphor, in the etymological sense of the term, because the *metaphora* is "that which takes you further." But this transposition is imperfect; the movement that she created was infinitely more rapid than that of Ulysses's ship or raft. The scale of the representation being known, the harmonization of the rhythms was impossible. Without the intervention of the gods and the unforeseen return of Ulysses, Penelope would have been condemned to yield to her suitors. She had nevertheless managed to make a *spatium* for herself, a space that was real between two uncertain days—but it is also a dreamed space, between two stops that would have brought Ulysses's vessel to Ithaca.

The paradox resides in the fact that she is captive in her own palace, which is a demure, a *demeure*, a *mora*. The *mora* is linked to time and not to space. The relationship between *demure* and its Latin etymology has been described by Zumthor: "It designated, in a timeless fashion, the fact of being there, in a sort of tasting of place."[1] Penelope is certainly there, but she is not savoring the place. The queen opens up a symbolic *space* through the help of her loom; she tirelessly weaves, but she endures the *place*. As for Ulysses . . . the King crisscrosses the seas, reaching the outer limits of the world and even sometimes crossing them. He covers immeasurable distances between points between which he is tossed about first with his men, then alone. He opens up a new space, on the margins of the Greek *ecoumene*, but he doesn't succeed in integrating this space in an intelligible geography. Monsters threaten him, as do his weaknesses as regards Calypso (a little), Circe (a lot), and Nausicca (quite nearly passionately). All along his trek, he is confronted by a series of points framed by immeasurable intervals. Sometimes, these intervals vanish, as the one that connects the island of the Phaeacians to Ithaca did. These points cannot be included in a superior geographic mechanism, in any logic whatsoever. Ulysses feels like a victim of *spatium*; he endures it—and for some he even makes fun of it.[2] He covets the return to place, his native island, whose *mora* recalls the principle of tasting, relishing. Ithaca has become a fantasy for him to the same extent as the *spatium* has become so for Penelope. The intimate geography of both spouses is the inverse, and at the same time, it reveals itself to be complementary. Penelope and Ulysses project themselves far away, one toward an open space where her husband is wandering about, the other toward the enclosed space of domestic comfort. It would take a while for them to meet. They would have to identify a common ground, a term that is not always pejorative. The breakthrough moment occurred in front of the trunk of an olive tree, which is itself rooted in the floor of the palace of Ithaca in guise of a private *omphalos*. It would be poetic to suppose that this landmark transformed into space for Penelope and into place for Ulysses. And as in the end the place is boring, Ulysses *deliberately* left for new spaces where even salt is unknown. I doubt that Ulysses was bored by Penelope; I prefer to think that he adopted his wife's point of view, which was infinitely more open and dynamic than his own. He should have chosen space rather than place.

The story of Penelope and Ulysses gives way to another, the one the women and men lived during the Middle Ages. The conception of the world could not indefinitely remain static, *local*, although sometimes otherness begins at home: in England, for example. Among amusing curiosities, Zumthor situates a writing of Gautier de Coincy, Prior of Saint-Médard de Soissons and author of miracles, which as it happens is a literary genre. According to Gautier, the popular belief that the English had tails was founded. The year is about 1230. To give his reader an idea of what the real distance separating France from England could be, Zumthor explained that it would take 16 days of travel by horse to go from Lyon to the other side of the Atlantic. Nevertheless, *space* was in the process of emerging. The presence of too many elements set the stage for an opening up to the concept of Elsewhere. The political *omphalos* was certainly located in Rome, but the imagined *omphalos*, Jerusalem, was shuffled to the margins of the world, at the threshold of the Orient, already in the Orient. The ideal center was placed at the border. This anomaly foreshadowed the willingness of extension, even expansion, that the Crusades revealed. It seems extraordinary to me that people began to "cross" each other the same year—or within one year at maximum—when the word *space* was forged. The beyond of this frontier both so new and so old was taking shape, even if it was often in the unbridled imagination of contemporaries. Oh, the throne of Prester John, so close to the Garden of Eden that the cartographers never forgot to reproduce it in their works. But the Orient, which resumed in itself this beyond, had also again become the destination of travelers. And when even these trips oscillated between the real and the imaginary—as that of Mandeville, certainly, but also that of Marco Polo—they contributed to removing the Middle Ages from the protective cocoon of Occidental Europe, from the *mora*, the faithful reflection of which was the innermost of the individual. According to Zumthor, "from the 8th to the 17th centuries, space extracted itself from the interior world of man, to become perfectly exteriority."[3]

It is at the beginning of this period that a radical otherness was conceived, similar to that of the Greeks during the fourth and fifth centuries BC. In both cases, the Other was Oriental. He was Persian for Aeschylus (in *Les Perses*, a tragedy dating to 470 BC) and for Herodotus. He was Arab and sometimes Persian for the crossings, and for all the numerous authors who embellished their precolonial gestures. Zumthor has delivered an analysis of this passage in

La Mesure du monde. He begins by establishing a classic opposition between here and beyond—a beyond that is situated "across," as indicated by the French term *contrée* or the German term *gegend*, substantives whose etymologies accentuate their adversative character. But, at this first stage, space is still a "neutral and pure space, indistinct, impermeable to sentiment as much as the gaze."[4] In my opinion, it's not even about a *space*; rather all *spatium* awakes the desire to fill that which is consubstantial. It is rather about a compact *place*, totally uniform even it its emptiness. Such a place takes its substance from a clear and absolute indifference and ignorance, in regard to what is located "across." The *contrée* doesn't start becoming a space until the moment in which it draws an interest, a modulation of the gaze. This process has been described with finesse by a geographer who was first and foremost a remarkable writer: Julien Gracq, the author of *Rivage des Syrtes* (1951), a novel whose temporal framework is not medieval, although the author mixes the epochs. "Across" from the little garrison of the Seigneurie d'Orsenna border post, where the young Aldo (the protagonist of the novel) is stationed, is Farghestan, which only takes on meaning in his eyes when he starts to acquire the geographic knowledge allowing him to transform place into *space*. In *Le Désert des Tartares* (1940), by Dino Buzzati, a book that is often compared to Gracq's novel, the lieutenant Giovanni Drogo adopted an analog posture. In the 1976 film adaptation of *Le Désert des Tartares* made by Valerio Zulini, we can see the Bastiani fortress, which would block out the northern march of an anonymous kingdom, that of the Arg-é Bam citadel in southeast Iran.[5] The landscapes that, in the film perhaps more so than in Buzzati's novel, deploy beyond the citadel evoke without doubt what the Orient "emptied" of the Middle Ages could have represented to the eyes of the outgoing *Christianitas*. The indistinctive *place* that the soldiers of the Bastiani fortress had long swept from their regard becomes, for its part, a *space* when horses appear at the horizon and begin to give it a body, opening the promise of a beyond, the stuff of dreams, but also the tangible object of a conquest to come. With Buzzati, just as with Gracq, two types of geography are designated, the principles of which Jacques Le Goff has established in an Italian article: the geography of nostalgia (which points to imaginary spaces projected beyond a closed world) and a geography of desire (completely attached to the willingness to "master the extent").[6]

It was in the Middle Ages, and we were just about *there*, when the territory that is the equivalent of *here* started to vibrate with curiosity and interest, and then to gape wide open at its margins. "Shortly," Zumthor continues, "the Occident would penetrate into a new mental universe where the contrary to presence would no longer be absence, but distance."[7] And the distance is the *spatium* between here and there, a distance that will be covered, then filled. In accordance with a logic that will establish itself bit by bit, that same logic that would inspire the geography of desire evoked by Le Goff, it is about measuring and mastering[8] the immensity, of measuring (*mensurare*) the immense, that which is *immensus*, immeasurable, even disproportionate, that which can still escape from the *mens*, which is the spirit of creation. The idea of a *spatium* that is a generator of space relates to the concept formalized by Plato is his work *Timaeus*. The latter distinguishes two types of meaning of place: the topos and the *chôra*. Adopting the perspective of a geographer, Augustin Berque has commented on this dichotomy and defined the topos as being a mappable place, the place where a body is located. This static approach to place is complimentary to that which can be seen in the *chôra*, the existential place proper to each thing—in other words, a seed or, as Jean-François Pradeau says in citing the geographer, the "affiliation of an extension."[9] "It is necessary," according to Berque, "to latch on to that which in the *chôra* can be a generating place; that is to say, an opening from which something can be deployed, and which justly does not limit or define itself."[10] It is simultaneously "print and matrix."[11] The *chôra*, as *spatium*, crystallizes a brief moment where place, perceived as a territory, prepares itself to move over into newness and the imponderable. The comfort of stable references is abandoned; a society abandons itself; it conserves the trace of that which is but accepts to become a matrix offered to the *im-mense*. It prepares itself to confront something else and someone else; it accepts to make *place* its own otherness, which up until then had not found the ability to express itself or to be expressed.

We are taken back in a sense to Greek cosmogony. Uranus, locked in *place*, is forced to retreat so that a new space that can emerge. Alas, the old myth stigmatizes from the outset that which characterizes the extensional approach inscribed in the history of the West. Although it is considered legitimate by the creators of the myth, it must be recognized that the overture is violent and disloyal: Cronus attacks Uranus, and blood is shed. Otherwise, the newness that is introduced by the son of Gaia—that is to say, the extension that he provokes—is relative. By a sad lack of imagination, Cronus tries to repeat a schema analogous to that conceived of by his father. He attempts to stifle the emergence of Zeus and his brothers and sisters. But he won't succeed, for the same reasons that enabled him to overcome Uranus. The myth (which is Greek) of the origins (which are Greek) of the world (which is Greek) nonetheless contains a particularity that is proper to the majority of cosmogonies: the Same and the Other are still confused. If the unconscious or unconfessed desire of the protagonists of the first narrative is to create something else based on the same, they strive to fill the gap of newness as soon as it allowed them to settle into the place of the Other, who is the father, in the depopulated landscape of cosmogony.

The challenge is to return to the minimal tradition that was established in the first generation, to the supposed comfort of home, a place seen as immutable, a model of stability. But this is impossible, because energy is already driving this embryo of society: history. According to the splendid image that Walter Benjamin recorded in his *Theses on the Philosophy of History*, in favor of a commentary on *Angelus Novus* by Paul Klee (which he acquired in 1921), history is this angel. Wide-eyed, open-mouthed, he observes the ruins that progress leaves behind. A strong wind blowing from Paradise pulls him away, still further, without him being able to fold his wings. In this way he reaches the future with his back turned, staring at a spectacle over which he has no control. His progressive and inexorable distancing demarcates space. And anyway, a return to the same is inconceivable: Uranus has definitively lost his attribute, his *genre*, a too-rigid identity. The myth doesn't approach the question of this loss, which is just as well a metamorphosis. But it is as if the Greek *mythos*, a so subtle journey, had at sometimes come back to the first random act of its convolutions. During his adventure in Lydia, next to Queen Omphale, Heracles embodied for three years the fate of Uranus. He too lost his attributes, but in a voluntary fashion in order take on others. In a way, history had made a leap, even if within a mythological landscape from which all historical depth is supposed to be absent. Uranus underwent the transformation without conceiving of an alternative, because that would have been even more inconceivable, while Hercules pretty much chose to test another gender, to test another identity. The nomadic hero had *spaced* his inscription in the world. The detour through transsexuality constituted for him (as well as for Zeus who put it into place?) a transgression, indispensable for the passage from a geography of place to a geography of space. This new geography supposes the emergence of an eventual otherness in the interval or gap of the *spatium*. It modified the perception of the world, making it so the gaze distanced itself from the *omphalos* and set itself to wandering far off, beyond place. This sudden change in perspective, whose genesis can be long—even yet there is still the question of *slowness*—is not reserved for gods, half gods, or even heroes. It is for all men, for all women, attached to a territory. It is even for the entire territory. For Gilles Deleuze and Félix Guattari, it corresponds to the moment when a "deterritorialization" is unleashed, which itself is inscribed in an eternal dynamic that is forever "passing-others" and "transgressive."

The Middle Ages did not escape this process of transgressiveness. What started to take place now had already happened in Greece. The conception of an *omphalos* instituting a vertical relationship with Heaven ended up by erasing itself for the benefit of a horizontal projection, translated by the opening of the heavens beyond the horizon, beyond the perceptible limits of an area gravitating around a center that itself seemed far off. The fantasy of an ideal and nearly inaccessible center opened the Christian world onto the idea of undefined space, which is virtually infinite. If Rome constituted the reasoned center of the *Christianitas*, Jerusalem represented the center of an ideal world that had to be built and first of all conquered. The task of making of its fantasized center the center of a territory on which it could exercise its sovereignty was left to Europe. Evidently, the awareness of this process was gradual. It was first necessary to probe the nature of the limits we had set and to be aware of the enigma that foreshadowed the horizon.

Gazing Far Away

We are strolling along on a beach. The probability of roaming around without seeing a soul is infinitesimal. But we are lucky: we see just a lonely walker, around whom a dog frolics. Seized by a Schopenhauerian impulse, we experience a sudden rise of sympathy for the quadruped. In certain seasons, the path along the beach reveals itself to be chaotic: isolation then becomes a lure. Walking along the beach will involve winding around rows of umbrellas and sandcastles on the shore, taking care to avoid the bodies of holidaymakers given over to the solitary yet collective pleasure of sunbathing, their navels in close communion with the sun god. Eyes chastely turned away from certain anatomical parts that are basking in the sun, one ponders the sacredness of this cult. But today, Mount Olympus is depopulated, although, among the few survivors, the titan Helios must surely remain. His beams have supplanted Zeus's lightning, which ceased falling ages ago. Our gaze is carried toward the open sea. The body of water is only slightly less crowded than the beach. We catch a glimpse of bathers playing with the waves, this or that watercraft tracing a dangerous and complicated loop, a far-off freighter, sails, sometimes shark fins (the sea can be filled with jaws). Although not completely clogged up, the horizon is still quite cluttered. This plethora upsets the desire for the distance. We can always strive to find a beach that escapes those waves of summer created and carried along by all kinds of motor vehicles. Does such a thing still exist in this wide world? According to travel agencies, no. In contemporary novels, yes. Jean Echenoz has identified such a beach in *Le Méridien de Greenwich* (1979), somewhere in the Pacific Ocean, to the south of the Midway islands: "The island is located in the center of the Pacific towards the Eastern edge of Micronesia, to the North-East of the Marshall Archipelago, equal distance between Shanghai and San Francisco, or for those who know them, between Ning Po and Eureka."[12] While we might not know it, we know that Ningbo, on the Hangzhou Bay, is across from Shanghai and that this particular Eureka, one of eight American Eurekas (we have done lots of research and have found them), is located to the north of San Francisco. We'll come back to this island a little later, but we have figured it out: access to it is not so easy. Anyhow, beaches don't seem deserted unless they encircle a deserted island, which is most often imaginary. Even in novels it is rare that beaches preserve their virginity. To indicate emptiness is to start filling it up. Finding the most suitable place for speculation on the horizon is a challenge.

At the time of high saturation, this emptiness might even inspire uneasiness. In our globalized society, the survival of the distant is an aberration. It would require the admission of failure in our system of full disclosure. It would point to a forgotten zone in the spatial-temporal compression of global perspective.[13] However, the unified landscape of a postmodern, even a posthuman, society sits uncomfortably with the idea of a horizon. The symbolic line that encloses the visible field is a type of scandal. How can we cohabitate with that which constitutes a limit to the visible? Antonio Prete poetically defines these confines in his *Trattato della lontananza* (2008): "The horizon is the line of the distance. It is the distance that represents itself and renders itself present while all the while remaining distant. It is the distant which shows itself in the form of border. A line where the visible touches the invisible. The visible seems accessible, and the invisible inaccessible; and with one like the other, the elsewhere is in relation. The horizon is the presence of somewhere else, the *mise en scène* of its possibility and at the same time of its exclusion."[14] However, the acceptance of the possibility of an elsewhere is not the marker of the contemporary. The modern beach is conceived to be crowded.

In former times it was otherwise. Because beaches have not always been overpopulated, at least not by sun-worshipping bathers. Prey to monsters, pirates, malaria, and "bad air," the beaches were abandoned to the poorest, to those who could not afford the luxury of experiencing fear. And to John, who in chapter 13 of Revelation stigmatizes the horror that could emerge from the depths of the sea, the sand, and the recesses of the soul: "And I stood upon the sand of the sea, and saw a beast rise up out of the sea, having seven heads and ten horns, and upon his horns ten crowns, and upon his heads the name of blasphemy."[15] Corneille echoes him in *Andromède* (1650), in which the heroine is promised to a sea monster that Perseus hastens to kill. On a coast adjacent to Sidon in Phoenicia, where she was playing with her companions, the nymph Europe saw the emergence of a white bull from the waves. It was Zeus, in disguise but still powerful, who ravished her. The beach was the site of abduction or of worrying spectacles. As Alain Corbin confirms, in a study that has become a classic on the perception of the beach and the sea between 1750 and 1840, "the ancient shoreline, as it is represented in the classical period, remains haunted by the possible emergence of a monster, by the brutal incursion of a foreigner, its equivalent; the natural place for unexpected violence, it is the privileged theater of abduction."[16] This shore aroused more anxiety before the classical age, before piracy was transported to exotic horizons, far enough away from Europe. Obviously, the beach had nothing recreational about it. And the sea that bordered it didn't have a better reputation. Alain Corbin reminds us that there is no sea in the Garden of Eden, and on the contrary, Genesis teaches us that the pouring out of liquid elements provoked the first and the greatest of catastrophes, the Great Flood. He who masters the floodwaters disposes of an incommensurable force. With the indispensable help of his God, because no human could appropriate such a power, Moses swallowed up Pharaoh's army in the waters of the Red Sea. And living far from the threat of floods became the tangible sign of Utopia. In the fourth eclogue of *Bucolics*, Virgil pronounced a Golden Age where ships, "nautical pines," would have ceased to transport passengers and goods, where the earth itself would have transported all things. Still in Revelation, in chapter 21, John had engaged in a similar prophecy: a new heaven and a new earth would be drawn, while the sea would have disappeared. But the sea, as well as the nautical pines, was still there. According to Corbin, we would have to wait until the middle of the eighteenth century for the perception of the beach and the sea to change, although it was still too far off at this time to apprehend the coast as our contemporaries do. The water should be cold or at least cool, salty, and turbulent. Southern beaches were banished. In the eighteenth century, Marbella numbered four thousand inhabitants and had introduced a novelty that was to revolutionize its economy: cane sugar. The Costa des Sol was not yet born. At the same time, Riccione had at most a thousand people and was developing an avant-garde project: to transform its beach into a rice paddy. Today, Riccione is one of the rare places on the planet where one is more likely to be stuck in a traffic jam at two o'clock in the morning rather than at two o'clock in the afternoon. What happened is that the Romagna Riviera arose out of the sand, and with it, a host of fashionable nightclubs.

"The liquid surface horizon in which the eye loses itself cannot integrate into the closed landscape of paradise. Wanting to penetrate the mysteries of the ocean is bordering on sacrilege, just as trying to pierce the unfathomable divine nature; St. Augustine, St. Ambrose and St. Basil were pleased to repeat it."[17] Corbin does so, too, in the initial chapter of his essay, in which he dissects the "roots of fear and of repulsion." Yet a few bold beings have tread on the beach and let their gaze drift across the surface of the water, up to the fragile limit of the horizon. But can we go beyond the horizon? Can we displace the horizon line? Among them, we could name Xerxes, who, from above on a mound where he placed his throne, was watching the Hellespont, our Dardanelles strait. But, if we believe Aeschylus (*The Persians*) and Herodotus (*The Histories*), he did not raise the fateful question. Perched at the top of the summit, the king of Persia did not contemplate the horizon: he followed the construction of a pontoon bridge that was to enable his troops to cross the strait and sweep over Greece.

As the water would not be tamed and the boat bridge broke up, the king had the whips brought out so that the waves could be beat three hundred times.

Afterward, he ordered a second attempt, which was successful. But that doesn't mean that he had mastered the liquid element. He had, on the contrary, committed an act of *hubris*: he tried to thwart the laws of nature to escape the restrictions imposed on all men, and therefore also on the kings of Persia. "A simple mortal, he imagined that he could beat all the gods, even Poseidon— the unwary,"[18] sighed the shadow of his father Darios, who returned from the beyond to comment on the folly of Xerxes.

It's Brendan that posed the fateful horizon question in the sixth century. Patron Saint of future navigators, Brendan, a native of the surroundings of Tralee, in Ireland, imagined that a ship could set off for the discovery of the Island of the Blessed. Tralee, which is located at the entrance of the Dingle peninsula, in Kerry County, is a predestined place for all those who wish to engage in meditation on the seashores. It is from this point that Brendan began to project himself beyond the limits of his island and his culture. The narrative of his exploits is recorded in *Navigatio Sancti Brendani*. The many manuscripts, which began to circulate in the centuries following the death of the abbot, did not testify to his willingness to go beyond the limits of the world. He would not be resigned to commit the sacrilege mentioned by Alain Corbin, as he was far from the pretension of "wanting to pierce the unfathomable divine nature." He would instead content himself with discovering Paradise on accident. Alas, he misplaced the coordinates at the end of a journey that itself followed the scansion of the liturgical calendar. He still accomplished a significant step forward in Western culture, though. Leaving the exclusive perimeter of the Mediterranean to spread out along the coast of what was not yet the Atlantic, the slowly upcoming Europe considered an option that was forbidden according to the consecration of the Judeo-Christian religion in parts of the Mediterranean basin: after all, why wouldn't waves have lapped against an insular Eden? This question brings us back indirectly to the vision that the Greeks, who, practicing *exokeanismos*, had projected onto their own Islands of the Blessed, beyond the columns of Hercules in the River Okeanos toward the West, there where the sun sets. The Vikings, they who would discover Iceland, Greenland, and Vinland (American?) at the turn of the tenth and eleventh centuries, were persuaded. In song XXVI of the *Inferno*, Dante also evoked the symbolic dimension of crossing the horizon, and of the entry into an "uninhabited world,"[19] "*mondo sanza gente.*" But he pairs any implementation of this geographical and ontological transgression with an exemplary punishment. Dante's modernism is extraordinary: his place in the Western cannon is enormous, as evidenced by the amount of criticism devoted to it. And what is more original is the growing number of novelists, often specializing in crime fiction, who are inspired by the *Divine Comedy*, such as Matilde Asensi (*Le Dernier Caton*, 2001), Nick Tosches (*Dans la main de Dante*, 2002), or Matthew Pearl (*Le Cercle de Dante*, 2003), to cite just a few.[20] But we sometimes tend to forget that the *sommo poeta* was a man of his time, at the beginning of the fourteenth century and that his vision of the world was not quite yet that of the Renaissance man, and even less that of a detective from the twenty-first century. For him, the horizon was what it has always been: a limit that man was not supposed to go beyond. Dante frequented Saint Brendan even more so than the navigators who were about to begin taking to the sea in his lifetime.

Philology offers a few precious teachings concerning this limit. It teaches us that the horizon and the city share an original boundary. The great urban planner Ildefonso Cerdá, who, in the second half of the 1850s, developed the new layout of Barcelona (a city whose expansion had been long blocked by constricting walls), was without doubt the first to make the link between the Roman plow and contemporary urban planning.[21] The *urbs*, of which Rome is the paragon, drew for certain authors[22] its etymology of handlebars from *l'aratrum* (the plow), itself named *urvum* or *urbum*. The *urbum* served to *urbare*, or to trace out a groove. For Cerdá, this act represented the quintessence of urban planning. With this linear view, Cerdá rather sees the symbolic origin of the city in the *urbs* rather than in the *civitas*. Cerdá illustrated this point with the plotting out of the Barcelona *Ensanche* (*Eixample*), a rigorously geometric checkerboard. To support his theory, the Catalan urban planner could have reinforced it by citing the *Aeneid*, and more precisely verse 755 of book V: "Meanwhile, Aeneas traced with the plow the city walls (*interea Aeneas urbem designat aratro*)."[23] By plotting the limit of Aceste (the future Segesta, Sicily), Aeneas yields to what will become Roman custom: the founder of a city uses a curved plow that displaces the earth inward from the traced surface; the furrow indicates the future walls; the gates correspond to the openings marked by the lifting of the plow (in order to let in the necessary staples and to remove corpses, as they could not be consecrated). Bringing together *urbs/urbum* and *aratrum* is practically a truism. The city is not turned toward the horizon; it cowers inside an enclosure destined to protect it and, no doubt, to set a limit to its pride. But this pride will prove to be without limit and measurement. According to Virgil, Alba was thus born and Lavinium too, founded by Ascanius, the son of Aeneas. Rome has a similar origin: it is worth recalling the story that attributes the tracing out by the sacred furrow (the *pomoerium*) to Romulus. It was necessary for Rome to see itself as *mobilis* and *augescens*,[24] that is to say, moving and growing, so that the production of the path (which is also the *lira*) becomes possible, so that a supposedly legitimate frenzy can take hold of it.

It was unlikely that Alba would open immediately onto the horizon, as legend attributed it to a Trojan sire. For Aeneas as for the Greeks, his former enemies, the horizon was a limit that the eye could not and should not cross. In Greek, *horos* was precisely a limit. What is *urbare* for the Romans was *horizein* for the Greeks: the action of tracing out a limit, a *horizon*. And again, temporality and spatiality intersect under the auspices of philology. The term *eschatology*, which refers to the study of the last and final matters of man, and by extension, the end of time, comes from the Greek *eschatiai*, the ultimate limit where the eye can see the boundaries of the known. Augustin Berque also corroborates the link between eschatology and *eschatiai*. To consolidate this link, he refers to Homer (*Odyssey*, I, 23) evoking *eschatoi andron*—that is to say, the Ethiopians, dark-skinned men, men coming from the world's end—and Herodotus (*The Histories* III) operating a full tour of the extreme zones of the known universe. Taking into consideration the dual metaphysical sense, which here includes a temporal and geographical valence, the horizon heralds the end of the world. To understand what it meant that Dante's Ulysses tumbled toward the unknown ocean, I think we should bear in mind the two meanings of this limit. Ulysses, the protagonist of the eighth *bolgia* (pouch) of the eighth circle of Hell, had indeed passed the Pillars of Hercules to wind up before the mountain of Purgatory, an island in the middle of the ocean without a name, the antithesis of the entrance to Hell. In return for his recklessness, which resembled a thirst for absolute knowledge, and therefore deemed him guilty in the eyes of Dante, he sank with his companions and completed his parable in hell, where he was transformed into a tongue of fire. Indeed, from the point of view of the Alighieri, his real sin was of a different nature. Ulysses had concocted the scheme to introduce Trojan Greeks hidden in the belly of a wooden horse. So it was he who had hastened the fall of the citadel, and it was he who still bore the responsibility of inflicting the pain of defeat and exile on Aeneas. But if Aeneas is to us the legendary founder of Italy, for Dante, he was the *historical* founder. In the eyes of the sharp Florentine, Ulysses was a fraudulent advisor to the "tongue of fire." Doubly fraudulent, even, as he should never have encouraged his companions to cross the Pillars of Hercules, the horizon of man, thus defying the law of God. Did he not know this god? Again: who cares? Ulysses was unforgivable.

Anyway, he would have failed in the limbo of Hell, in *nobile castello*, in the company of all those who had not been baptized. He would have then found Homer, who Dante hadn't read because his models were Latin (Stace, in particular). The medieval Ulysses was not the precursor to great navigators. In one essay of the *Nine Essays on Dante* (1982), Jorge Luis Borges examined song 26 of the *Inferno* and completed a critical overview of the journey of Ulysses in its Dantean version. In a postscript concluding his text, he added the following clarification: "It is said that Dante's Ulysses foreshadowed the famous explorers who, centuries later, would reach the coasts of America and India. Centuries before writing *The Divine Comedy*, this type of human being already existed. Erik the Red discovered the island of Greenland around the year 985; at the beginning of the 11th century, his son Leif landed in Canada. Dante could not know that. The Scandinavian tends to be secret, to be like a dream."[25] Dante had not read Homer; he hadn't read Adam of Bremen either, who, in his *Gesta Hammaburgensis Ecclesiae Pontificum*, a 1075 chronicle that retraces the history of the archbishop of Bremen and of Hamburg, from which came the Nordic world, reported the discovery of Vinland. But Adam had also reported on a "Land of Women" located somewhere in southern Finland, a country populated by Nordic Amazons. Was he serious? In fact, Adam had made a small mistake: he had mistranslated the toponym Kvenland, whose meaning remains obscure. Still today, *kvinna* means "woman" in Swedish. But "Kvenland" isn't "Kvinland." The emergence of legends sometimes comes from nothing, such as the simple and innocent change of a vowel.

If the Land of Women existed, if it existed today, it's a safe bet that it would lie on the edge of the masculine world. That is at least what Canadian novelist Aritha Van Herk seems to think. In *Places Far from Ellesmere* (1990), she imagines a citizen of Alberta deciding to spend her vacation on Ellesmere Island, the tenth largest island in the world by its extension, but also one of the least densely populated. On the cover of the novel, which is presented as a *geograficione*,[26] the island takes on the form of a curvy feminine body whose edges are covered with names of places, as in the chorographies of yore. The relationship between text and place has manifested itself, much more so than at the time of Dante. And the heroine of the novel will push this logic to its limit. She isolated herself in Ellesmere in the company of a novel that should enable her to cope with sleepless nights in northern Canada. She could have chosen *Crime and Punishment*, which exalts the white nights of St. Petersburg, but she opted for another big (and grand) Russian novel, *Anna Karenina*. In the fleeting tranquility of the Arctic summer, she began to annotate the book, the heroine of which suddenly starts to oscillate between Russian snow and Canadian ice. For Aritha Van Herk, Anna Karenina is a kind of female Ulysses, eager for new knowledge. She heads off on the discovery of a type of "beyond femininity," pent up by codes kept under the hostile eyes of the demons of society, who, in Moscow as elsewhere, have replaced the gods of Olympus. Her exploration takes place through reading. However, as noted by the Canadian vacationer, who writes in the second person, "the day you return from the glacier, you realize that Anna is condemned because she reads."[27] And, facing the horizon, this feeling is accurate. At the top, in Ellesmere, the horizon is subjected to "a strange phenomenon: in a certain light and at certain moments it flattens out to the point that you are able to see beyond, above, and even behind it. Everything is imminent. It's a trap of light and of latitude, here, in the northern extremity. Reading is done via an overhanging. The possibility of engaging in an overhang arises for any reading, just as there is the possibility of jumping over the words to see what happened before it happened. That is what Anna does: she reads herself by jumping in front of herself."[28] And she died from it. As Dante's Ulysses, she learned too much, she was too bold. It is definitely not a good idea to be the pioneer who crosses the horizon. For Anna, justly (or unjustly), the line of the horizon and the lines of the text succeed the lines of the railroad tracks that bring the train that crushes her.

But let's leave Tolstoy and his Albertan exegete there. Let's return to Dante who was contemporary to those who, in Genoa, had preceded Columbus exactly two centuries in his attempt to reach India by the west. As land routes from the East were less accessible due to Muslim control of Christian holy sites, a detour by the sea was envisioned: the circumnavigation of Africa. In 1291, Tedisio d'Oria (or Doria, the name of a large family of Genoese sailors and shipowners) decided to fund the first long-term shipping expedition outside the familiar waters of the Mediterranean. Ugolino and Vadino Vivaldi, commanding a crew of three hundred men divided into two galleys, embarked on a journey that was to last ten years. But they lost body and soul, probably in southern Morocco or off the shores of Senegal after—who knows?—landing in the Archipelago of the Canary Islands. No one was ever able to explain what happened to the small fleet of Balearic sailors. Another Genoese, Lancilotto Malocello, officially discovered (or rediscovered) the Canaries in 1312. Lanzarote owes its name to him. Anyway, the shipwreck of the Vivaldi brothers was predictable: the traditional galleys could not resist the swell of the ocean (which the Spanish Armada would confirm at its own cost in 1587); they were forced to make frequent stops. What did the Vivaldis do during these stops, before they sank? I would like to think they trampled along the sandy beaches of Africa or the Canary Islands in search of a beyond that moved at the same pace as themselves.

First used in France and in French, the *horizon* made its entry into the European vocabulary after 1250. It was an *orizonte*[29] at the time. According to Michel Collot,[30] in the astronomical domain, the term indicated both the "sensitive" horizon (the circular line that seems to separate heaven and earth) and the "rational" horizon (the circle passing through the center of the earth that divides it in two). But the term has long been uncommon. Collot states that Marc-Antoine Muret, a commentator of Ronsard's *Amours*, chose to gloss "When the sun" (sonnet LVIX) in 1553; this poem referred to the "Horizon," relying on the authority of Cicero.[31] In book II of *On Divination*, the great Roman orator had translated the Greek etymon by the word *finiens*. *Orizonte* didn't definitely become *horizon* until the seventeenth century, at the price of an etymological repair. Dante was the first great writer to use the word (*orizzonte*) in his *Purgatory*. But this horizon didn't yet open onto anything in particular. It fed what some commentators have called a "poetry of astronomy."[32] The first verse of song II, which includes the term, effectively introduced two *terzine* where the Florentine strives to locate Purgatory in relationship to Jerusalem: the horizon is the meridian that connects one to the other. Opposite of the Holy City, the mountain of Purgatory was in the southern hemisphere, uninhabited by the living, according to Dante. For the Vivaldi brothers, the concept didn't even exist yet. They saw the pure *spatium* completely ajar before them, a huge gap. Few others have shared the same sensation that they must have felt.

All history books site the opening of Europe onto the Atlantic Ocean, a *space* of discovery *par excellence*. But few are those who wonder about the meaning of the words they use.[33] But what about *ocean*? and *Atlantic*? It's the Greek's River Okeanos that had gradually metamorphosed into the ocean. In the fourteenth century, the great sea Ocean that surrounds the world was still being evoked (by Mandeville, among others). But the epithet eventually supplanted the noun. The sea stopped being the ocean, it became the ocean. And it is this ocean that became the Atlantic. The innovation would first intervene in English. Ranulph Higden, a Benedictine monk who led a long and undoubtedly peaceful life in a monastery in Chester, wrote the *Polychronicon* in Latin, an ambitious chronicle whose subject was the history of the world from Creation until 1357. Ranulph did know that the world was even older than one could imagine at his time. He wrote his narrative in seven books—seven, just like the number of days required by God in Genesis. In the first book of the chronicle, he mentions the *oceanus Atlanticus*, which coincided with the waters washing on the Occidental shores of North Africa, the land of the giant Atlas. Using image and text together, like many before him, he elaborated a great map in which the world was surrounded by a river. The *Oceanus Atlanticus* was little more than a hamlet, somewhere off the cost of Mauretania, the land of the Moors. Thirty years after the story was interrupted, bringing us to 1387, John Trevisa, chaplain of Lord Berkeley the fourth, undertook the translation into English of the *Polychronicon* and coined the expression *ocean of Athlant*. For the record, we can note that if translating into English is now a common activity, at the end of the fourteenth century, it was otherwise. According to several linguists, Trevisa was one of three scholars (the other two being John of Cornwall and Richard Pencrych) who saved the endangered English language. English didn't die, and the Atlantic was born.[34] Trevisa was decidedly a great man.

The Atlantic name covered the western maritime boundaries of the known world, the limits that were thresholds that could be crossed nimbly, either to explore the African coasts just below or to see what was beyond the horizon, offshore, there where the sun was setting. We waited still later for the Atlantic Ocean to become what it is for us today. In fact, we had to wait until geography and cartography incorporated new data on the wider world. The Atlantic began to take shape after the first major transoceanic voyages and circumnavigations of Africa. For cartographer Sebastian Münster, author of *Cosmographia*

Universalis, toward the middle of the sixteenth century, an *oceanus Occidentalis* arose to the north of the equator and a *mare æthiopicus*[35] to the south. For Mercator, the *oceanus* was *atlanticus* in the north, and the *mare aethiopicus* became an *oceanus aethiopicus*. Shortly before, in his *Theatrum Orbis Terrarum* (1570), Ortelius had made of *Atlanticus* a universal ocean. Münster, Mercator, and Ortelius were three of the principle cartographers of the sixteenth century. There were others, notably the Spanish, for whom the Atlantic was the *Mar del Norte* and the Pacific *Mar del Sur*. Here, a brief excursion to Nicaragua is required, where two villages are both called by the docile name of San Juan. To distinguish them, the idea was to call one San Juan del Sur and the other San Juan del Norte. San Juan del Sur is located to the north of San Juan del Norte. Despite appearances, there is nothing strange here, because the first overlooks the Pacific and the second the Atlantic. It was not until Varenius's *Geographia Generalis* (1650) that the equivalence between the Atlantic Ocean and the Mar del Norte was established—quite a bad idea after all, since later the Atlantic would be separated from the North Sea. But this is a different story, if I may say so. In any case, the two San Juans one day opened their doors to Americans, but in diametrically opposite circumstances: that of Sur (which is to the north) to Mark Twain, who visited it in 1866; then Norte (which is to the south) to the Marines of Ronald Reagan, who believed that the electoral defeat of the party he supported was incompatible with the principle of democracy. That principle is still good, as we know.

One day, Nicaragua had an American president. Would he win Reagan's favor? It's not sure. Patrick Deville discussed this "ridiculous and sublime"[36] fate in a story called *Pura vida. Vie & mort de William Walker* (2004). William Walker, a native Tennessean, arrived in Central America in 1855, after having tried in vain to establish a southern republic in Baja California, around Sonora. Later, with the help of some mercenaries, more precisely "twelve thousand buccaneers [*filibusteros*] from North America,"[37] as soberly stated in the 1898 *Diccionario enciclopedico hispano-americano*, he seized power in Nicaragua. It was not long before he was proclaimed president. That was in July 1856. In May 1857, defeated by cholera, mass desertions, and the opposition of many, he returned to the United States. Instead of giving up, he again defied fate by storming Honduras. He was executed in September 1860, arms outstretched on a beach, just outside of Trujillo. "He seemed resigned to his sad end. He was shot," adds the *Diccionario Hispano-americano*, in its article on "a certain [*un tal*] Walker." This Walker makes us irretrievably think of Aguirre, the "Wrath of God," a magnificent and pathetic conquistador played by Klaus Kinski in Werner Herzog's film, which was inspired by Lope de Aguirre's Amazonian journey in 1560. Walker is part of a lineage of men who love, above all, to fight a hostile environment in order to overcome it. What a vain and ultimately silly project. At no point did Deville imagine that Walker was able to stop on a beach to contemplate the horizon. Perhaps it is for this reason that his exploits were never anything other than pathetic, criminal raids. His views lacked grandeur and especially the kind of measure that portrays the spectacle of the impassable distance line, that line that traces out the horizon.

In his book, Deville speaks of another man facing the sea: Simón Bolívar, who had planted a cane chair in the sand on a Colombian beach, shortly before his death at the end of 1830. The writer quotes a letter that the *Libertador* addressed to General Juan José Flores, on November 9 of the same year. In this letter, he said, "He who serves a revolution plows the sea."[38] He added that emigration was the only sensible resolution in America and that the country (Venezuela) would fall into the hands of cruel tyrants—Venezuela, like Nicaragua, like so many others, at random within an eventful history. The *Libertador* was tired, disillusioned. He surmised that his great work would not survive him, that it would be quickly forgotten. The mobility of the ocean waters increased its perplexity. These waters devoured everything. Did Bolivar know that the horizon he contemplated was born out of the separation of Gaia and Uranus, that he was the furrow that traced the original tear? The culture of the ancient world interested him less than that of the New World. I've already spoken of Cronus and Zeus, his son. I might add that, according to Hesiod, the union of Gaia and Uranus was also the outcome of Mnemosyne, Memory, which in the words of Antonio Prete, "is image, narrative, rhythm. That which preserves time against the dying of time."[39] Obviously, Mnemosyne took Bolivar under his wing, since we are still talking about him, and often at that. What a nice allegory, so classic! But the horizon remained open. The "moribund Pharaoh" continued to dream in the sand: "Behind the horizon, he imagines the main island of Cuba that he would like to snatch away from Spain. Therein lies his most terrible military regret."[40] And so he leaves the horizon, which so often creates greed. And greed itself leads to a dream of conquest, a desire for violence. Maybe there are counterexamples of this so-Western dynamic that has not left Latin America unscathed. Maybe we should make the effort to look in other places, other than in Europe or the Americas, for a horizon that indicates that "in every limit, the beyond-the-limit trembles."[41] "Tremble," yes, because except for the arrogance of the conqueror, indecision reigns.

The Western Horizons

It is an exceptional moment when the statuses of geographical spaces set themselves to wavering. Then there is the hesitation between several qualifications, including the traditional division between "real" and "imaginary," which can no longer be made. This moment of indecision, which is sometimes prolonged, is that which precedes the journey of discovery. Then the day comes the gaze is drawn to the horizon and projects the individual toward a beyond that it is up to he himself to shape. The ocean, river or sea, has long constituted a totally imaginary space, a space subject to unbridled imagination. It would be futile to try to establish here the history of the place that the liquid mass, extending beyond the Pillars of Hercules, occupied among the ancients. There had been, without a doubt, "real" navigations. A famous journey would have brought the Carthaginian Hanno, who continued the Phoenician tradition, as far as the coast of sub-Saharan Africa and perhaps even to the current Cameroon. Conversely, the Greek Pytheas of Massalia would have traced back to the north, near Thule, the northern limit of imagination taken to the extreme. As stated by Predrag Matvejevi,

> if the waterways did not let themselves be easily identified, it may be because they were intertwined with stories: the maps on which they are indicated could be imagined, the writings that accompany could also be invented. Historians and geographers have struggled against Pytheas: Strabo did not believe that he had succeeded where 'the Tropic of Cancer is the Arctic Circle,' and where the ground is such, 'that it is impossible to walk or to navigate,' just as Polybius saw his travels as mere tales . . . There is a limit, affirmed the wise, between the probable and the improbable, or, as one would be inclined to say these days, between the figures of history and forms of narrative: that limit, the great adventures have gone beyond it and surpassed it.[42]

Before Hannon, before Pytheas, around 600 BC, if we believe Herodotus (*Survey*, IV, 42), Pharaoh Necho ordered Phoenician sailors to sail along the coast of Africa (Libya) from the Erythraean sea (the Red Sea), to go around it, to exceed the Pillars of Hercules, and to sail back to Egypt via the Mediterranean. And Herodotus comments: "They brought back a fact that I find incredible, if others add faith to it: in bypassing Libya, they said, they had the sun on their right."[43] After all, if we are only a little less suspicious of Herodotus, we estimate that Pharaoh Necho's Phoenicians just got two thousand years ahead of the chronology established in the West. However, what Herodotus presents as an absurd fact (the sun on the right) is the strongest argument in favor of the reality of crossing the equator and this pioneering circumnavigation. In another travelogue, which is commonly called the *Périple de la mer Érythrée*, written in Greek in the first half of the first century AD, sailors and traders have crossed the waters to India (this is evidenced by the accuracy of the descriptions), or even to China, because it refers to a place called *Thina* where silk was produced.

If the hypothesis was verifiable—but it is not—Columbus was preceded a millennium and a half by sailors aboard sailing skiffs much frailer than his beautiful caravels. It is not just a dreamed geography or the geography of hypothesis that the Phoenicians, Carthaginians and Greeks had once nourished. As their exploits would have minimized the performance of navigators of medieval Europe, which had become a crucible from the Western point of view, the temptation to cast doubt on their discoveries was irresistible. It was essential that Western Europe inaugurate the discovery of the world in order to better take possession of it. Truth be told, known geography was itself exposed to mistakes. Being poor navigators and not too curious about what lay beyond the limits of the *Mare Nostrum*, the Romans incorrectly perpetuated their Atlantic knowledge. In the early Middle Ages, they had forgotten the existence of the Canary Islands, the virtual incarnation of a place beyond Greece, although Pliny the Elder had mentioned it in one of the volumes of his *Natural History*. Only the Arabs remembered it. This omission was for a time a source of happiness for the Guanches, aboriginals from the Canary Islands. With few exceptions, the gestating Europe neglected the idea of crossing the ocean horizon. Fortunately, we retort, if we think of the fate that was reserved for the Guanches, figures who, before others, possessed an otherness that would be savagely reduced. Among the infringements on this intellectual enclosure, there was the chronicle of Adam of Bremen, about which I've already said a few words. There were other speculations, sometimes late. In his voluminous *Navigazioni e Viaggi*, the publication of which began before his death in 1557 and continued thereafter, the Venetian humanist Giovanni Battista Ramusio (of whom Manlio Brusatin says he "gathered the extraordinary stories of shipwrecked travelers who, after having experienced what it was like to touch the ends of the earth (and life), managed to return to their homelands"[44]), evoked the journey of the brothers Nicolò and Antonio Zeno to the wide open West. Let's accompany them briefly.

In 1380, Messer Nicolò the elder sailed "towards the north wind," that is to say, to northern Europe. Off the coast of Flanders, his boat was struck by a storm that took him to the coast of Friesland island, in the south of Iceland—that is *if* we believe the evidence of a map published in 1558 by Nicolò Zeno the Younger, a descendant of Antonio. Despite its name, Friesland does not have any connection to the Frisian Islands. Friesland was also larger than the Ireland of St. Brendan, the great precursor. And Friesland was far away from everything that was familiar to the men of the time. Messer Nicolò sailed in the "occean of Athlant" that John Trevisa was in the process of introducing into a vocabulary reserved for a few scholars. A sovereign called Zichmni ruled Friesland. He welcomed Messer Nicolò with open arms when he learned that he was Venetian. Shortly after he settled in, Nicolò sent a letter to his brother Antonio, asking him to join him as soon as possible, because he knew about his brother's desire to see the world. The Zeno brothers prolonged their stay for several years. After a series of skirmishes with the Norwegians and their Icelandic allies, Nicolò died. His brother Antonio took over. Master navigator, he was sent by Zichmni toward the west, where a fisherman claimed to have failed in prosperous islands. In a letter to Carlo, another of his brothers who remained in Italy, Antonio announced the discovery of Estotiland, located over a thousand miles to the west of Friesland. The island of Estotiland "has, in abundance, all of the world's goods, and is a little smaller than Iceland, but more fertile, it has in its center a very high mountain from which spring four rivers that irrigate it."[45] And now the earthly Paradise and its four rivers had themselves also started to drift westward! The island was populated by people who did not understand any known language but who owned a few Latin works.[46] Back in Friesland, Messer Antonio told about his exploits. Zichmni then decided to sail with his Venetian guest: "And so we set off to the west."[47] The small fleet sailed beyond Estotiland, and after defying "the darkest of seas (*il più cupo pelago*[48])," they landed at Icaria, which was named after the first king, who himself was the son of Daedalus, King of Scotland. As Zichmni insidiously tried to capture the land, he provoked a reaction on the part of the Icarians that pushed the Frislandais away, offshore. They continued their journey westward, where they discovered an unknown and seemingly deserted island that they called Engrovilande. Zichmni decided to establish a small colony that certainly did not worry the rare indigenous inhabitants spotted by the Frislandais. Even more, they lived secluded in caves. But the crew eventually faced a forced wintering. The king adventurer remained on site and ordered Messer Antonio to take command of those of his men who wanted to return to their homeland. In a final letter of which, according to Ramusio, there remained only the first paragraph, Antonio told Carlo that he had written an account of his travels, but, as then lamented by the Venetian humanist, this book was hopelessly lost.

Many have commented on the correspondence between Antonio and Carlo Zeno, which was then reproduced by Nicolò Zeno the Younger and by Giovanni Battista Ramusio. For some, Zichmni could have been Henry Sinclair, Earl of Orkney, who lived in the late fourteenth century. The Scottish nobleman not only landed in Greenland, but he also discovered Nova Scotia (1398). He would, in fact, have discovered America a little more than a century before Columbus. Some put forward what is deemed an irrefutable argument: "American" plants are carved inside Rosslyn Chapel, the chapel of his fief whose construction was completed in 1486. Others believe that they are just stylized plants or even vulgar raspberries. In any event, the Canadian island was officially invested by John Cabot, another Venetian, in 1497 and later by the French and the British. As for the Mi'kmaq who had been living there quite a while, they must have thought that Zichmni, alias Sinclair or Cabot, was Tweedledee and Tweedledum. For all practical purposes, I would also like to mention that Nicolò Zeno the Elder was tried for corruption in Venice in 1394, when he was supposed to be in Friesland. Regarding the status of Friesland, it is recognized that everyone was simultaneously right and wrong. Indecision is coextensive with the transformation of space into place and at the entrance of the latter into objective geography. At this early stage, it is impossible to distinguish between the real and the imaginary. A law professor whose course I once followed was fond of saying that a door was open or it was closed. Between literature and geography, the door is always ajar. Friesland and the small islands encircling it never existed other than in stories and maps (up until the eighteenth century for some). But perhaps they covered the area of the Faroe Islands, which, in all truth, had been known about since the ninth century. These islands were Friesland without quite being the Faroes. In turn, the Faroe Islands owe something to Friesland. But what? If the place does not hold onto at least part of the fiction, it is doomed to sink into boredom. And sinking is such a shame, for an island.

The Sublime Dive

From within the interior of lands, it is rare for the visible horizon to be clear; it is usually obstructed by human artifacts or higher ground. However, for an individual standing on a beach, looking out onto the sea or ocean, the horizon is usually free, except during the month of August, when there are the ripples of swimmers and other surfers at this beginning of the twenty-first century. We can calculate the distance between an observer and the curved line of the horizon with the utmost precision using the following formula: on a clear day, regardless of atmospheric refraction, this line is at a distance, D (in kilometers), equal to $3.57\sqrt{h}$, h indicates the height (in meters) at which the eyes of the observer are with respect to a zero elevation. If his eyes are placed five feet two inches (1m 60 cm) from the ground, at sea level, this individual will perceive a horizon at a distant of four-and-a-half kilometers. If the observer were a little shorter, the horizon would be a league away, which is a more poetic measurement. It seems that this league is the exact interval that separates the known from the unknown. The landscape unfolds between the observer and the slightly curved line that separates the sky from the crest of the waves. However, as noted by Michel Collot, whose corpus of work surveys the romantic horizon and the horizon of poets, "the landscape is perceived as an extension of personal space, its scale measures the magnitude of one's own body enlarged to the limits of the horizon."[49] If the horizon were fixed, the span of this fantasized projection of the body would almost be negligible. It would take climbing Mont Blanc for the landscape to really stretch out, for the horizon to enlarge, or, to borrow another formula from Collot, to know "the inexhaustible reserve of untapped opportunities."[50] At the summit of Mont Blanc, according to the official arithmetic, the horizon opens up onto 248 kilometers. Perhaps it is for this reason that the Romanticists have symbolically claimed residence on the peaks. It was not just about gaining height; it was also necessary to take some distance and get perspective, in order to create the possible. But whether one is located a little more than four kilometers from the line, or a little less than 250 kilometers, the ultimate observation is the same: the end of the world is near. In addition, access to the summit of Everest being somewhat difficult, it would not be easy to push this limit. But as nature would have it, when it is not disturbed by man, the horizon is not fixed. An imaginary line that moves at the discretion of the observer's point of view, the horizon was intended to supply dreams of the distant, of the far-off, of the beyond. This is what, as pointed out by Collot, is so essentially *fabulous* about the horizon: the horizon emphasizes the poetic disposition of space, in fact, "a fable of itself [that] lends itself to poetic fabrication."[51]

In *Nudités* (2009), Giorgio Agamben discusses a range of issues of critical importance to creation and salvation and what we cannot do on the glorious body or in the last chapter of the world's history. He also wondered about contemporary nature and its problematic anchoring in current society. According to Agamben, to be contemporary, the poet must fix his gaze on time, on *his* time, and accept to perceive darkness rather than light: "contemporary is he who receives the beam of darkness that comes from his time full in the face."[52] It remains to be understood what "'see the darkness', 'perceive the darkness'"[53] means. A priori, it would be an anonymous action, for darkness, as we say colloquially, does not talk to anyone. Based on so-called exact sciences, and more specifically astrophysics, Agamben teases out the beginning of an explanation. We learn that in our expanding universe, the most distant galaxies are moving away so quickly that their light cannot enlighten us. In fact, these galaxies are moving away at a speed greater than that of light. In other words, darkness is a light that does not reach us. Agamben then arrives at this wonderful conclusion: "To detect in the darkness of the present this light which seeks to join us but yet cannot, that is it, to be contemporary. That is why contemporaries are rare. This is also why being contemporary is, above all, a matter of courage: because it means being able to not only to take a look at the darkness of an epoch, but also to perceive in this darkness a light which, directed towards us, is infinitely going away. Or yet still: being punctual to appointments that cannot be missed."[54]

The contemporary is basically the one who accepts that the star flies in the opposite direction and that humanity, faced with its many limits, fails to receive its light. He knows that the star is absent and that darkness is the remnant of light. He is not a *tenebrio*, a "friend of darkness." He simply accepts the idea of waiting, to endorse the growing gap that separates him from the light that was. Let there be light . . . and already it was as if the God of Genesis had resolved himself to veer off at the speed of distant galaxies. Once upon a time, the Roman augur issued a similar view. For him, the star was *sidus* (*sideris* in the genitive form). The dictionary confirms that the word was neuter. But it isn't quite completely neuter: sideration as speechlessness cannot be neuter. For the augur, when the sky was empty, and the absence of the stars frustrated expectations, it marked the *de-siderium*, another grammatically neuter word that sanctioned the absence of any favorable signs. Of course, the *desiderium* brought the disappointment or the regret one feels when facing emptiness. Moreover, in Latin, *desiderare* was both "to desire" and "to deplore." In favor of an appointment, which Agamben tells us he will miss, the contemporary feels a doomed desire to remain unsatisfied. Of course, there is a close link between this fleeting light and the horizon that recedes as one moves toward it. Before one as the other, the desire to fill a void is expressed.

In the early fifth century BC, an artist whose identity escapes us painted a fresco inside a stone sarcophagus, near Paestum (Poseidonia for the Greeks), in Campania. This fresco, discovered in June 1968, represents a diver jumping into the emptiness below from on top of a group of columns. It is splendid, and this fresco makes of its discreet author one of the greatest artists of all time. Who is the diver? It is impossible to know; it is even impossible to interpret with certainty his act and the environment in which he was present. The mysterious mural has inspired abundant comments, very few of which agree. Pascal Quignard, who devoted a story to Butes, one of the Argonauts (the one who, unlike Ulysses and his companions, plunged into the sea to join the Sirens and their music) has formulated an alternative:

> Either this man who dives is a young man who is pushed by the crowd of the stone acropolis of Poseidonia, head first, his penis dangling below his belly, unexcited, arms outstretched, flying in the white air before touching the sea water where the crowd threw him. Or, this man who dives is any dead man the moment when, arriving at the confines of the living world, taking his momentum with his feet posed on the Pillars of Hercules, he plunges into the underworld represented by the green water of the Ocean and the tree with the leaves of Oblivion.[55]

Making the diver into a sacrificial victim associated the scene with punishment or an ordeal of some sort. This reading does not satisfy me. I prefer by far the second, which makes of the columns in the mural the Pillars of Hercules. One question is to know what extends beyond this symbolic limit. The waves that bathe the foot of the columns on the western side are probably not those of the River Okeanos. They evoke death and an afterlife. But perhaps they are the waters of the River Okeanos, which in themselves require a mortuary interpretation. The "beyond" of the geographical world would then also be a biological "beyond" or better, something "beyond" metaphysics. One thing is nevertheless certain: The diver has based himself on the columns that mark the limits of the world in order to plunge himself *plus ultra* in a style bordering on perfection. What was he contemplating from the top of the columns just before diving? Was the man entirely given over to the emotion that shook his body and his mind? Had he obtained inner peace? Did his eyes still have the capacity to fix on an exterior spectacle? If it is the case, his gaze could have only been drawn toward the horizon. But then what became of the boundary indicated by the columns, the *finiens* of Cicero? A provisional limit, relatively. Beyond the limit unfolded a new limit that indicated or distinguished the horizon. The diver of Paestum is a beautiful incarnation of the *de-siderium*, this inextricable mixture of desire and lamentation. In any case, he is on time to an appointment that he is sure to miss. He strives to materialize the absolute limit by diving, but at the top of his platform, if indeed he lifted his head, he has seen the line of the horizon, a new limit that has shown him that any limit is a threshold that indefinitely expands space. Perhaps it is this sublime awareness that has attributed too much grace to this ultimate dive. Two trees frame the diver and the columns. Off on an invisible island, the tree is vigorous; its branches stretch upward. Below the columns, the tree is stunted; a branch seems broken. The space beyond is vital. Maybe it indicates the Isles of the Blessed or the garden of the Hesperides. In any case, for the Greeks it was unknowable. In many of his works, Pindar mentioned the existence of the Pillars of Hercules. According to the Hellenists, he was the first to do so. And his warning was clear: "After them, the path is inaccessible to superior men as well as to vulgar men. I do not aspire to follow beyond: I would have to be crazy!"[56] Pindar was nearly a contemporary of the author of the fresco that decorates the tomb of Paestum.

Dante at the Beach

In his splendor, the diver contemplating the horizon really distinguishes himself from ancient times. In the Middle Ages, the presence of the horizon incited moderate enthusiasm. Just like the landscape, it does not appear in the list of what we would call today, in tourist vocabulary, the major *points of interest*. Technically speaking, we could imagine a person might stay on the beach to watch the spectacle of the horizon, but, to the extent that this initiative would be part of a subjectivity that had not yet been asserted, it would have seemed confusing to the others. However, what was frankly inconceivable was that an individual would attempt to physically get closer to the line, to displace it, using an offshore movement leading to the Ocean sea, to the unknown. Let's return for a moment to Dante and Ulysses, who later became an avatar of the diver in the Paestum fresco. The Florentine wouldn't dare have imagined that a man could materially reach the horizon and become the contemporary of his own desire. The fulfillment of this eccentric desire would have threatened the supremacy of God. As recalled by Michel de Certeau, "throughout the Middle Ages, and even in the 16th century, it was accepted that morality and religion *have the same* source: the reference to a single God holds together a historical revelation and a cosmological order."[57] For Dante, the configuration of the universe was still stable; nobody could challenge the cosmological order. Not even Ulysses. The division was simple. Dante allowed himself to think that the light of God's celestial body was visible in Heaven and that darkness accumulated in Hell. When Cato of Utica, guardian of Purgatory, received Dante, he was surprised that he managed to "emerg[e] from the depth of night / That makes the infernal valley ever black."[58] Darkness was not yet another form of light. In fact, what is the color of the horizon? As a poet, Antonio Prete has attempted to answer this question in his *Trattato della lontananza*: "At sea, the horizon is a line where all shades of blue and of heaven alternate, intersect, and overlap. If the line of the sea horizon is to the east, the rising sun abolishes the border, gradually dissipates the darkness, transforming sea and sky into a glare that spreads towards the day. If the line of the horizon is to the west, fire and explosion of colors, as they go out, announce the evening, and with the night, the disappearance of the horizon."[59] For Dante, the show is different. The beach of Purgatory is bathed in a light that the pilgrim has finally found. The air has the "sweet hue of Eastern sapphire."[60] The clear air is more colorful than it is usually. But the horizon? At night, we could see it as "Mars with fiery beam / Glares down in west / over the ocean floor."[61] In the highly accurate calendar of the poet, we are at the dawn of Easter Sunday, April 10, 1300. And the horizon is illuminated by a bright light, the light kindled by the Angel ferryman in his passage, the angel who leads souls into the intermediate kingdom.

For Dante, as we have seen, Ulysses had sinned. He was subjected to the violence of an unusual desire. He had invested unknown space and therefore *desideratus*, a space deprived of stars. He should not have come close to the island of Purgatory, whose "solitary shore . . . / That never sailing on its waters saw / Man, that could after measure back his course."[62] Dante the character (and the author) still sees the sea as a source of isolation and danger. In Purgatory, there is no beach to speak of. The coast, battered by the waves, "on the oozy bed / Produces store of reeds. No other plant, / Cov'd with leaves, or harden'd in its stalk / There lives, not bending to the water's sway."[63] The bank is not the place of a centrifugal reverie; it is blocked from the waves by a barrier of reeds. And yet, the reed is a symbol of humility as, on a metaphorical level, the reed of Pascal could have been. But Dante was the first man alive—or, more accurately, the first man able to *survive* the trip—to see the sea and the sky of southern antipodes. He discerns "Four stars ne'er seen before save by the ken /

of our first parents,"[64] those of Adam and Eve, because let's not forget that the Earthly Paradise crowns the mountain of Purgatory. For him, the place is not *desideratus*. It is sidereal. It is also amazing.[65] As in Hell, the pilgrim is stunned. According to the extraordinary geography of Dante, there is only one way to reach the antipodes and Purgatory, which also corresponds to the only moment when the eye is free to wander away from the sea. We learn that the soul of the deceased is transported "by the shore / Where Tyber's wave grows salt,"[66] to await the Angel ferryman. And as explained by Casella, a deceased Florentine musician whom Dante meets in song II of Purgatory, "for there always throng / All such as not to Archeron descend."[67] The pier of Purgatory is at the mouth of the Tiber, thus at Fiumicino, which a few centuries later becomes the location of the main airport of Rome. Wings and more wings . . . The path taken by Casella and the other pending souls is the only one that leads to Purgatory. Only two living beings have been exceptions to the rule: Dante, who, following his journey from hell, crossed the globe from one side to the other to land on the beach of the "second region (*secondo regno*),"[68] where he put his foot down, and Ulysses, who made a U-turn in the world before falling to the sacred mountain, with his experienced crew.

In *Horizon mobile*, Daniele Del Giudice has the impression that Antarctica was once exiled to the end of the world, torn from the land and warmer climates. And now on the ice yet other wings, although truncated, start to show: "There was a conviction and a sigh that only those unconscious, surreal penguins were kept as angels."[69] In the ice desert, the narrator raises a question that would appeal to any reader of Dante, and especially an Italian reader whose worldview is more or less shaped by *Lectio Dantis*: "And I wondered how Dante could understand that purgatory was here, there where he placed it, exactly below the southern sky."[70] Could the answer be prosaic? Let's be iconoclastic enough to not exclude it. If Dante discerned with great acuteness the ontological or eschatological implications of his act, he ignores the cosmographic scope. Thus he avoids scanning the horizon. When he does, it is to contemplate the arrival of the Angel ferryman. That's it. Anyway, he knows he is himself an exception. No other man is expected in these parts. This posture is obviously linked to the symbolism that orients the architecture—rather than the geography—of the afterlife of Dante. The pilgrim in love compares himself to a dumbstruck surveyor in one of the last verses of the *Comedy*.[71] Everything is arranged in a vertical axis, whose evolution occurs over descents and ascents of the hero and, for a time, his guide Virgil. Let's move away from Lucifer and the infernal funnel; let's approach Beatrice and Paradise by climbing up the cornices of Purgatory. Dante looked up. He often describes the starry sky, which he measures with the skills of an astronomer. He invested in a divine plan. He never fixes the horizontal gaze of the human world, whose presence in these ultramundane latitudes should be abolished. *Should*, I say, because it is transiently expressed by the intrusion of Ulysses. The horizon is not divine geography, but it is just as closed to human desire. The horizon is a sign of pure *space*, yet Dante evolves in a *place* saturated with various connotations that requires no alternative. From there, it is impossible to go elsewhere. In his trip, being the only one registered in a horizontal dimension, Ulysses strayed from the general dynamic. According to Yuri Lotman, who wrote some pages on this journey in *Univers de l'esprit* (1966), "Ulysses travels as if he were located on a map."[72] As for Dante, he *peregrinates*. Lotman then adds that Ulysses "becomes, for Dante, the Renaissance man, the first discoverer and the traveler" and that "this image appeals to Dante because of his integrity and his strength, but he is also repugnant to him due to his moral indifference."[73] I do not believe that Ulysses embodied the Renaissance man for Dante. That would be something of an anachronism. No doubt he embodied, in a less problematic way, the one who subverted the scheme governing the medieval world. Ulysses was both the Other of Dante and another Dante. In his otherness, the Greek appeared to the Florentine, a fantasized descendant of the Trojans, a being that was both reprehensible and admirable. Facing the horizon, Dante still adopts the attitude of the medieval man. For him, the line does not delimit two parts of the world, one of which is visible and real, and the other virtual and waiting to be discovered. Rather it marks the boundary between heaven and earth. The horizon of *sommo poeta* is still oriented according to the vertical. But the author is also the first to have used the word *horizon* and the first character to have contemplated it—without ever scrutinizing it. The horizon is there, a mile from the beach in Purgatory. But the space that rotates around it is not yet opening. This is the genius of Dante: to have pointed it out, almost reluctantly. The border was not yet a threshold. It was closing in on something rather than opening up on something else. Perhaps it was easier to try to discover the divine beyond rather than the human beyond.

Well after Dante had evoked the three dimensions of the afterlife, the Spanish writer Alejandro Gándara imagined a trajectory of a middle-distance runner in a beautiful novel called *La media distancia* (1984). Between two training sessions, Charro, the hero, receives an explanation from his friend Vidal as to what is a vanishing point: "There is, over there, in the distance, a point beyond which you cannot continue. The earth continues, the plain, the trees too, but neither you nor I can see them." When Charro asked why he called it a "vanishing point," Vidal replied, "Because if I could overcome it, it is not what I see that takes me away, but what lies beyond. I would run away from here, I'd go, I'd become another."[74] He would go to America, his dream. By the force of things, America is not yet a dream for Dante. And the vanishing point is not at the end of the horizon but first in the depths of the infernal pit and then in the heights of the mountain of Purgatory. Besides, Dante does not seek to escape: he strives to set himself in place. He is not even in the throes of the spatial drive that will haunt generations to come and lead them to overcome all the *non plus ultra* of this world and other worlds to come.

Navigators are affected in the long term, but so are those for whom the horizon embodies a sublime desire. Let's return briefly to the *Rivage des Syrtes*, so that these geocritical dynamics are discussed from beginning to end. In his own way, Aldo find himself at the Admiralty, where he was sent to a type of modern Purgatory beach. He himself employs the term, because it is usually in this remote place that officials of the Lordship of Orsenna atone for some mysterious error in their service. During an excursion to Sagra, in the vicinity of the Admiralty, Aldo sees "the reeds with hard stems which are called the 'ilve' blue" and notes that "no clearing had begun on the deprived land."[75] It sounds like Dante describing the shore of Purgatory Island. But if the Florentine pilgrim only mentions the horizon in passing, even though he had done so as a pioneer, Gracq and his narrator repeat the word fifty times or so. I found 47 occurrences in the novel, and it is possible that I overlooked some. Aldo continues to direct his gaze toward a beyond that the horizon seems to block. "You want to see something rise up from this empty horizon,"[76] Marino, his supervisor, remarked. Marino never reproached Dante this way: he would never have had the opportunity to do so. It is true that for Dante, the island completes the wildest expectations of the being: the top of the mountain, which stands in the middle, is crowned by the earthly paradise. Again, the vision of Dante is local (linked to location) and nonspatial (due to space). From his point of view, the early beyond materializes spiritually upright, as we have seen. As for Aldo, it is striking that desire is not considered in any other way than as an "omnipotent horizontal." He himself develops his own mountain of Purgatory, but a little further, just beyond the horizon: there is the Tangri volcano, whose name refers to the divine sky of the Mongolian people and which overlooks the entire Farghestan coast. Unlike Ulysses, Aldo managed to avoid having a wreck while "cruising" in territorial Farghestan waters, which led to the foot of the mountain. Without doubt, in the twentieth century, the yearning to give concrete form to desire is less objectionable than at the time of Dante. It is in any case what is encouraged by the maxim of the Aldobrandis, a major family of the Lordship of Orsenna: *Fines transcendam*. And it is this maxim that Aldo implicitly adopts.

From the moment we considered the fabulous horizon from a beach, we began to turn the page of the great book of the history of mentalities. Dante turned the shore of Purgatory into an unloading dock for the Angel ferryman and the souls waiting to be redeemed. Others just after him have made the coast a jetty leading into the distance. This has opened a new perspective as elusive as the horizon line, one as fragile as its edge. They had to understand, in an intuition bordering on the unlikely and for us what is almost indescribable, that the elusive horizon was like the darkness that heralded a distant light. It was a new act of faith. In sharing, they will finish by understanding that this imagined light, never before seen, beyond the darkness and the horizon, was something other than the trace of a divine presence out of reach forever. In its own way, medieval theology placed the source of this light vertically to the celestial spheres. Dante's *Commedia* embodies the finest representation of this projection. It was not so much the persistence of galaxies expanding that Agamben mentioned in *Nudités*. After all, it is placed in a down-to-earth dynamic that would challenge the limits of an exclusively human, or perhaps even too human, world. But if this trend was down-to-earth, it was also *sublime*, as it was going to burst the limits of the horizon, turning the last frontier of the visible world into a threshold of the unknown. In the darkness, in the apparent void of light, what symbolized the beyond of the horizon would break free from *space*. It is in taking the risk of confronting the Sea of Darkness, another name for our Atlantic, that one could take place in time, in a terribly current time, present-ified. Crossing the horizon, roaming the seas in search of a light that would otherwise be ignored, was the task assigned to those who wanted to be the contemporaries of a new world that would in turn result in the discovery of a New World.

CHAPTER 3

The Spatial Urge

The Myth and the Void

It took a great and adventurous spirit to pierce the mystery of the horizon so that geography could stop being relegated to confined space and parataxis. Who was the first to take an eager look beyond the visible? Who turned his attention to the sublime, which opens up an area where the limit (*limes*) is likely to become a passable threshold (*limen*)? Very clever is he who can respond. In addition, how can we isolate the ancient or even antique response to this modern question without falling into overinterpretation, without falling into cultural anachronism? In cosmogonies, the myth speaks the origin of the world. It does not evoke the amazement of beings facing blank space, whose identification requires a subjective and pioneering perception. And this perception will be broken down across the mode of internal focalization.

However, let's try to think of an example that illustrates an early epiphany about space. The first episode from antiquity that comes to my mind is clear: it is the crossing of the Straits of the Symplegades[1] that the different versions of the Argonauts' journey reproduce consistently. Here, chronology is floating. Greek cosmography loved associating the extension of time and the expansion of geographic areas with maritime exploration. Sail the ship, and the universe is built. The Argonauts' voyage is particular in that it includes the telling of the first official Greek navigation, and it is also a generation ahead of the voyage of Agamemnon and his men to Troy. Therefore, it precedes by another ten years the wanderings of Ulysses through a Mediterranean that was not yet the Middle Sea. Jason sailed for Colchis, whose capital Aia (or Ea) "is located at the extreme limits of the sea and the earth."[2] His gaze directed toward the rising sun, surrounded by a crew of fifty heroes from all over Greece, he accomplished a feat whose implications we cannot necessarily measure. Commander of the very first ship, the vessel Argo, he discovered the essence of navigation. To reach the country where the Golden Fleece is preserved, he and his men must furthermore change seas, and with that, change universes. According to the geography of the time, this passage materializes in the Symplegades Crossing, which was still called the Cyanean Rocks, the Planets, or the Clashing Rocks. For today's reader, the location is known as the Bosporus, which is usually considered as a *locus amoenus*. For the Argonauts, it is a stone monster endowed with life. A real monster, in fact. Overcoming the obstacle is like triumphing over the goalkeeper of the beyond, in both the metaphysical and geographical senses of the term. But we know that these two epithets are always an almost inseparable couple.

Unlike the *Odyssey*, the text of which was soon to be established under the uncertain and prestigious auspices of Homer, the *Argonautica* have seen several versions, among which no author has reached the very highest levels of the literary canon.[3] The best known of these is undoubtedly Apollonius of Rhodes, who directed the Library of Alexandria in the third century BC. In his *Argonautica*, Jason and his men deliver Phineas from the clutches of the evil Harpies who, day after day, ruin his table and condemn him to die of starvation. To reward Jason for his generosity, Phineas prophesies the stopovers to come. In particularly, he teaches Jason how to escape the Symplegades, "because they are not sitting on deep roots."[4] The Argonauts are to set out a dove. If the bird is able to overcome the constricted passage, then the men can depend more on the strength of their arms than prayers,[5] and they should row hard and firm; if the bird gets caught, the men should turn around, because even if the Argo is made of iron, it will not be able to resist the embrace of the monster. The Argonauts comply. The jaws of stone gape half-open. Euphemus launches the dove, Tiphys the pilot gets going, and the rowers bend the spines on their oars, trembling. The situation is critical, but Hera was watching over Jason, like Athena watched over Ulysses. Her great protection allows the vessel to pass. On the other side, the crew discovers that which no Greek has ever seen: "The heroes had to take a deep breath after the terror which had frozen them; they lifted their eyes to the sky and the open sea, which carried on out of sight: they said to themselves that surely they had escaped from Hades."[6] Jason finds that the most formidable obstacle has been crossed. He might have added that the Argonauts have experienced the epiphany of pure space. The next episode is less conducive to surprise. Through his account, Phineas provides a path made of place names; through the announcement of her protection, Hera levels doubts; one and the other fill the space through anticipation. The Argonauts would have to wait for their painful return and the trip around the world (which would be asked of them) to plunge again into the unknown, into smooth space. In any event, the busy sea changed valence after their initial navigation. And the toponymy bears the trace of this inflection. The waves of what is for us the Black Sea had become hospitable to the Greeks. The sea, the *pontos*, had suddenly become *euxeinos*, although it had been *axeinos* up to then. We can turn yet again to Predrag Matvejević, who explains this semantic evolution: "The Black Sea (*Pontos Euxeinos*) has an unusual etymology. The adjective *euxeinos*, which means hospitable, replaced its opposite *axeinos* (inhospitable), since that is how it was shown to those who went to conquer the Golden Fleece; perhaps it is a popular etymology, which would have transformed the meaning of the old Persian word *akseana* (dark, black): the north was marked by black."[7]

Valerius Flaccus, whose bibliography is incomplete for us, designed a new version of the Argonaut journey in the early seventies AD. This decade was not easy. It framed the destruction of Jerusalem and its temple by the Roman legions and the eruption of Vesuvius, which sealed forever the fate of Pompeii. If the author is less prestigious than Apollonius, who alongside Virgil was his model, his Argonauts have become even more famous. Now they have in their ranks Ulysses and Aeneas, who put to the test their future nautical prowess. But, whatever the composition of the crew, the obstacles remain: the Symplegades continue to stand in the way. The strait is boiling over more than ever. This is a type of basin that has been shaken too hard and that could overflow everywhere. At the top of Mount Olympus, the gods are nevertheless vigilant. Somehow, Jason and his crew pass and the sides of the strait are stabilized. But the heroes, fearful of forcing the passage again in order to return, ignore it. Here they are; it's their turn in a new world: "At this point the waves on which nobody had sailed for a long succession of ages were struck with amazement at the sight of the ship. All the earth of the Bridge became flat and motionless, all its kingdoms and far-off tribes become accessible. Nowhere else does the shore fall away before the advance of the sea."[8] And Valerius Flaccus keeps alive the description of people and places that the sailor heroes will meet in the sea that has been rendered navigable. In ignorance of these parts, the prophecy of Phineas once again fades. However, once out from the Symplegades, the Argonauts of the Roman writer have also tasted a moment of pure space. As with Apollonius, this epiphany occurred during a lull. The heroes realize that they just escaped death; they don't yet feel the need to organize the next stages of their trip. They enjoy an interval of down time, a pure relaxation that is a priori incompatible with their status—a hero is not supposed to remain inactive. Industrious, ingenious, he must constantly project himself over time, albeit broken up by unexpected events, and a series of different locations. The degree of freedom that he is able to make for himself in this absorbing system—it's not what the hero wants! It's too small. It does not have the luxury of prolonging its time; by necessity, it is momentary. And it is precisely in this fraction of time that space appears, smooth, the ridge seeming a foreign idea, beyond the reach of any standard. Even in the myth, this moment is not going to last. Perhaps it is too demanding, maybe it is irreconcilable with the need for action that drives the individual.

There is ecstasy, ex-stasis: it leads the one feeling it to a different movement, to a projection toward the sublime. The epiphany of pure space marks the entrance into a possible world—an entrance that one is not always ready to take.

The link has often been made between the foray into a new geographic area and a sudden burst in the metaphysical beyond. The evolution takes place in an ambivalent context marked by the dangerous, even mortuary, dimension, in which is placed an undertaking that exceeds human limits and that falls within the supernatural. But we also discover the prospect of a new life, or at least a life that is otherwise extended, in a world committed to expanding the human experience. The spectral connotation of the Argonauts' voyage has not escaped readers' and critics' attention. It seemed to accentuate over the centuries and versions of the story, to the point of becoming manifest in the *Argonautiques Orphiques*, the latest transposition of the story from antiquity: a beautiful, anonymous text. But the different authors have invented nothing: the interpenetration of the worlds began from the *Odyssey*. Developing this point would take too long here, just as it would to reconcile the different *Argonauticas* and the odyssey of Dante's Ulysses as well. How tempting it would be to compare the crossing of the Symplegades with that of the Pillars of Hercules! Suffice it to remember that, in Homer, Circe urges Ulysses to go to Hades in order to consult the shadow of Theban Tiresias on the trip home. Ulysses will not have to steer; it suffices for him and his men to be carried away by the blast of Boreas until they reach the "end of the ocean."[9] There, Ulysses must go to a rock at the confluence of the infernal rivers, which indicates the entrance of Hades. After shedding tears of fear, they depart. In the space of a day, at a rate that contrasts completely with the pace of the *Odyssey*, they reach "the edge of the deep ocean." We arrive in the city of the Cimmerians, in the north, "in the mist and vapors,"[10] there where the sun ceases to shine its rays (it is by this region of outer limits, which became more material in the fifth or sixth century, that the orphic Argonauts will, in turn, make the return trip). The return trip of Ulysses's Greeks will be just as fast, or even more so, as the trip out. In one night, they return to Circe's island, at the horizon where the sun set on the world. For Ulysses, the north and west of the universe are only separated by a day of navigation at the most, a briefness explained by the benevolence of the gods. We must not forget that in Greek geography, the world was stretched in its width, and the north was crammed onto the *omphalos*. Therefore, Epirus, the northern region, was *epéiros*, the "mainland," the rest of which would be Europe to us, though it was practically unknown to the Greeks of Homer. We mustn't forget that there is at least another voyage whose rapidity exceeds expectations: that of the Phaeacians taking Ulysses back to his island of Ithaca before being punished by Poseidon.

Hades opens at the confluence of the infernal rivers. Space, for its part, opens at the intersection of worlds, one new and one old, one real and one mythical, one material and the other a dream (or nightmare). The spectacle of space remains intermittent. Too fast, it is worrisome. In *Medea*, Seneca focuses the action of the play in Corinth; the drama unravels, ending by putting Jason into opposition with the Colchin princess in the eyes of King Creon. A priori, the Symplegades are at a safe distance, both in regards to space and time. But, through the intermediary of a chorus who seems to sure know a lot, Seneca returns to the memorable feat of the Argonauts. Certainly they dared. Certainly, they innovated by inventing new navigation techniques that the playwright delights in describing. But more important, they broke the balance established in the Golden Age. They tore man from a serene enclosed place, where he was allowed to age in peace, to plunge him into the depths of open, gaping space: "But a world divided into a balance found itself unified by the Thessaly vessel, which submitted to its blows the surface of the water, and enclosed by our own torments the seas which remained far away."[11] As a reward for their performance, the Argonauts received the Golden Fleece and Medea, a worse curse, according to the chorus, than anything else. And Seneca denounced the compression of the world, a type of globalization before its time: "The Indian drinks the fresh water of the Araxes, the Persian that of the Elbe and the Rhine."[12] Of course, Seneca says here that the world of the Indian and of the Persian borders now that of the Armenian and the Germain. But we cannot help but think today of another liquid that this Indian, this Persian would drink: fizzy, sweet, brownish, produced in Atlanta. Universal. The real thing. According to the chorus, which returns twice to what it presents as an act of *hubris*, the Argonauts "atoned by a cruel death for violating the laws of the sea."[13] Fifty lines are devoted to recalling the doom of Orpheus and some of these early sailors. Jason's fate is clear. Moreover, Medea is given to him. One thing is certain, in the eyes of the chorus and, no doubt, those of Seneca: after being forced open, the world will never close again because "later in the course of the years, a time will come when the Ocean will release its grip on the world, when the earth will open up in its immensity, when Thetis will reveal new worlds and Thule will no longer be the limit of the universe."[14] The prophecy of Seneca was well worth those of Phineas and Tiresias. It is true that Seneca knew a lot about the end of the world. He was born in Cordoba, in an Andalusia that was still part of the *Hispania Baetica*, and that, to the west, was closer to the western limits of the Roman world. Although he left Cordoba young, he had the opportunity to smell the spray of the River Okeanos, beyond the pillars.

Epiphanies of Space

Seneca died at an acceptable age for his time. But his death was premature. We know that Nero ordered his days to be cut short. If he had been immortal, Seneca could have kept his nose to the wind in the company of Columbus. Can we imagine the two men engaged in peripatetic debate somewhere on a beach in *Hispania Baetica* or Andalusia (which is the same)? Would they talk about stoicism? Would they have speculated on what would have forced the ocean, river or sea, to loosen its grip on the world, to make a *place* for (Western) man? Would they have considered the negative consequences of a presence that would quickly become pervasive? If Columbus had just been two centuries older, it is Dante that would have summoned him to the vestibule of his Inferno. After all, "moral Seneca"[15] stood not so far from Orpheus, an Argonaut at times, the geographer Ptolemy, or even Averroes, "who made that commentary vast"[16] of Aristotle. It would take a little imagination to visualize the conclave of the Florentine and the Genoese. Perhaps a little less to sketch the silhouette of Christopher Columbus striding forward with impatience and seriousness on the sandy shore of Palos de la Frontera or the lava coast of La Gomera, the small Canary island that was the last bastion known on the road to the New world. The sea or ocean, or the sea ocean in the vocabulary of Columbus, is open to all projections. Normally, large bodies of water function to reinforce continuity, an interval between the lands. As Predrag Matvejevi wisely reminds us, "the Greeks had several words for sea: *hals*, meaning the salt, the sea as matter; *pelagos*, a body of water, the sea as vision or spectacle; *pontos*, the sea as both space and route; *thalassa*, a general concept (of unknown origin, Cretan, perhaps), the sea as an experience or event; *kolpos* meaning breast or bosom, and designating the maritime space that embraces the shore: gulf or bay."[17] Did Columbus's caravels scour the *pelagos*? Certainly, the spectacle was daily, though not always exciting. The *thalassa*? Yes, because the event was significant, even if it was not the one we thought: it was far, America . . . But the three Spanish caravels especially took the *pontos*, just like one would take a bridge. Because, far from accepting the rift between worlds, it is up to the Genoese navigator to plug the hole, to fill the gap between the lands. And this bridging action does not correspond to a ratification of space. On a larger scale, somewhat magnified, it restores the vision of the Greeks, for whom the original *pontos* constituted a safer liaison between the islands, the surest developer of the *archipelagos*.

On August 9, 1492, in his diary, the navigator noted that the gentlemen of the island of Hierro (the most westerly of the Canary Islands) and La Gomera had affirmed, some even under oath, to see land offshore toward the west. Columbus had received similar reports on Madeira and the Azores. The extreme island of the archipelago was never seen as a final ground. Something had to be found beyond there. And this assumption was necessary for the completion of Columbus's project. It was necessary for there to be lands to the west, as well as birds, grass, or reeds in the seawater to foreshadow them. The narrative of the crossing betrays moments of wonder. On October 8, the admiral noticed that the air was as sweet as it is in Seville in April, and it was fragrant. But four days later, the discovery of "America," the outpost of which was the Bahamian island of Guanahani (Watling, for us), was the subject of a story that leaves the modern reader unsatisfied. The proximity of a land, which was moreover inhabited, was marked by a stick floating on the water that, according to the men of *La Pinta*, had been worked by the hand of man. This clue and others filled the crew with joy. But we cannot rely on the logbook to reflect the sentiment of Columbus when facing the new space! The concern lies elsewhere, for example, at the top of the mast where the lookout was perched. It had been planned that the first to see land would be rewarded. Who would receive the annual pension of ten thousand *maravedís* and the silk tunic that Columbus had promised to add to the kitty? In the night of October 11 to 12, two hours after midnight, the honor fell, it seems, to Rodrigo de Tiana, a sailor from *La Pinta*, which was in front of the flagship. But the idea was maintained that the first sight of land should be held by Columbus. Before anyone else, he claimed to have discerned, before witnesses, lights flashing in the distance, about ten o'clock on the evening of October 11. He preferred to keep quiet, allowing time to check his fleeting impression. The bonus was rightfully his then, if we may say. In the early morning, the admiral went ashore aboard a boat, accompanied by Martín Alonso Pinzón and Vicente Yáñez, captain of the *La Niña*. He displayed the royal flag, embroidered with an *F* and an *I* (the initials of Ferdinand and Isabella the Catholic) and a cross, and took formal possession of the place. The upcoming program, which was very clear, was announced. Columbus mentioned, all the same, naked men, very green trees, fresh water, and fruit. This place was still an open space.

In *Message* (1934), Fernando Pessoa evoked, as one would a ghost, the metaphysical empire that the Portuguese navigators had once traced far beyond the familiar waters. One of the poems in the collection is called *Horizon*. It is not Columbus he speaks about, as Pessoa grants exclusivity to his compatriots, and even if the Genoese Columbus could be Portuguese, he would eventually become Spanish. However, the horizon is stateless. Without doubt it would be more accurate to say that it is irrelevant to the concept of homeland. As for Pessoa's poem, he recites the various facets of the sight that unfolded before the sailor's gaze at the approach of virgin land. First there was the "severe line of the distant coast,"[18] which fulfilled an expectation. We know how Columbus longed to see this line; we can guess the relief he felt at its sight. But then "the earth, closer, in sounds and colors unfolds: / Finally, when we arrived, there were birds, flowers, / There where far was nothing but an abstract line."[19] Too far ahead of the time of the Romantics and their ability to reproduce landscapes and sensibilities, too eager to reign in his emotions under the austere framework of an official account, or even a statement about his travels, the discoverer only imperfectly recounted the scope of his wonder. However, he gave form to an abstract line, that of the horizon, that of the imagined, far-off coast. He didn't take long to include his discovery, using rectangles or rhombuses, on a map. Between these two moments, he felt the thrill of space; he allowed himself to be penetrated by the novelty; he was suspended in something that had not yet been transformed into a place that you control. But it was on the tip of his tongue that he experienced this feeling: just, it was not worthy enough to be communicated. The hero did not let himself go to ecstasy. It was up to him to control an environment of which he was an integral part. The description of the space corresponded to a breach in modesty, to a timid confession, freshly formulated. Columbus was like Jason. The first sailing is ideally renewed each time that the thrill of space propagates. For the record, Columbus knew Jason so well that he identified with him. In the *Livre des prophéties*, which he wrote between 1502 and 1504 and in which he placed his mission under the enlightenment of the Scriptures, but also of some classics, he quoted the verses of Seneca cited here: those that announced the opening of the world. But he and Jason had another thing in common: one sought a fleece, the other sought nuggets, both of which were made of gold. For many commentators, the different *Argonautica* were the putting into poetry of the first Greek expedition directed toward the gold deposits of Caucasia. By a subtle alchemical operation, the stone marking the *omphalos* was turned, it seems, into a philosopher's stone that transformed the substance of limits into gold. This transformation was called a "projection." Later, Mercator's "projection" concretized the alchemical projection in an unexpected form. The *Atlas* is the plan, programmed by the West, to occupy the land, and its projection is the safest way to find gold. The *Atlas*, or the map of Treasure Island, was sought by both Jason and Columbus.

Far from any coast, in the heart of the steppe where the short grass seemed infinite, did Marco Polo experience an epiphany of space? In *Marco Polo* (1982), the beautiful, romantic transposition of *Devisement du monde*, Maria Bellonci has repeatedly returned to the emotion that the merchant would have felt about the wide world opening before him. He must have experienced "the reality of pearl divers, sailors and merchants from all countries involved in the bold challenge which was launched in space."[20] And the hero himself exclaimed, "Under my eyes opened the map of infinite journeys."[21] Shortly before making the return journey to Venice, the Romanesque Marco Polo recalls the years spent in a world that, without being completely new, was opening wide. He is conscious of having experienced "the intuition of the great outdoors" and thus having "cured and satisfied his attraction to the unknowable."[22] But the Marco Polo of the twentieth century is not the Marco Polo of the late thirteenth century, who was much more discreet about the feeling that was inspired in him by having access to the vast expanses of Central Asia. Was there more modesty in the expression of subjectivity at that time? Perhaps there was a lack of appropriate tools to give shape to the delight inspired by the incredible, to this intuition of the great outdoors evoked by the character of Maria Bellonci. But the latter, I repeat, was equipped with a culture and a mind-set that does not belong to the medieval model from which one of its avatars comes.

Ibn Battūta, sometimes wandering about in the same steppes as Marco Polo, sometimes on the edge of the desert where there are dunes after dunes after dunes, sometimes on the deck of the boat leading to distant shores, was he more forthcoming about the sensation he felt regarding the opening of space? The fact is that the man from Tangier did not have the occasion to marvel very frequently. Although his trip had included an exceptional extension, it was mainly to take him on a tour of the Muslim world. Logically, it was not possible to invoke otherness within a space subject to a common religion. It so happened, however, that Ibn Battūta went beyond the limits of the referential universe. So he complained. In his presentation of the *Voyages* of Ibn Battūta, Stéphane Yerasimos evokes the figure of Ibn Fadlan, another Arab traveler. Ambassador of the Caliph of Baghdad, Ibn Fadlan first went, in 923, to the northern steppes, the Dasht'i-Qiptchaq, with the king of the Bulgarians. His mission failed, but he recorded his travel souvenirs in *Voyage chez les Bulgares de la Volga*. For Ibn Fadlan, the new space unfolding before his eyes raised a considerable, even insoluble, problem: the days lengthened indefinitely, the light upset the usual rhythm and the normal course of things. The Bulgarian king's tailor, from Baghdad and a Muslim, suffered especially: "And he said that for fear of missing the morning prayer, he dared not fall asleep at night for a month. Someone who puts the pot on the fire in the late evening makes his morning prayer before it boils."[23] The eye focuses on the poor demonstration of a pot that is boiling at the wrong time. The tailor and the ambassador are as confused as Ovid once was, he having been banished to not far away from there, at Tomis. Ovid was initially frightened by an environment that seemed hostile. Ibn Battūta probably did not set foot in this country, which, in and of itself, did not prevent him from making a description of it. But, when it came to the Land of Darkness,[24] he expressed a fear similar to that of the Roman poet.

Columbus's gaze was greedy. As for that of some of his subordinates, who happened to resent the hierarchical relationship, it was eager. Gold fever had already broken out. Perhaps, when all is said and done, neither Polo, nor Ibn Battūta, nor Columbus experienced the thrill of *space*. We have even less access to their deep feelings as none of them directly wrote his own story.

As Predrag Matvejević noted, "those who put the deepest passion in travel and navigation do not always have the time to note where they went and what they saw: the fact of traveling for them is more important than the story."[25] Accuracy is imbued with wisdom. Polo dictated his adventure to Rustichello; in Granada, Ibn Battūta gave the details of his trip to a certain Ibn Juzay; and as for the original logbook, *Journal*, of Columbus, it was lost very quickly and the text that we do have (published for the first time in 1825), was taken from a copy by Bartolomé de Las Casas. Most of the records contained in the *Journal* are reported in the third person. About Columbus, Matvejević adds, "His companion during the second crossing, Juan de la Cosa, wrote more and better than him."[26] In fact, Juan de la Cosa, who also took part in the maiden voyage, was an especially better cartographer than the Genoese. In 1500, he drew the first map of the world that included the discoveries of the previous great decade.

Perhaps the intuition of space was purest with an anonymous person. Has there been a person in history who is unknown and who was visionary enough to have prompted a group to undertake a migration beyond the familiar, beyond the measured, beyond the apprehensible? Who could say? Who could say precisely who was behind the first settlement of the "Americas," the community of naked men that Columbus told about on the beach in the Bahamas? The French say of someone who too swiftly claims an innovation or a discovery that he "beat down an open door." The Italians have a similar expression (*sfondare una porta aperta*). But they can also say about someone who is audacious and naïve that he "invented hot water" (*inventare acqua calda*). I understand the French expression; but I confess to a lesser understanding of the Italian expression, which seems deeply unfair to the anonymous genius. Inventing hot water means that you have mastered the first spark, which is not nothing—it is something that is extraordinary. Opening the horizon has undoubtedly constituted an act as crucial as the consequences of having invented hot water. The horizon has never been an open door. It has never been beaten down because it is constantly hidden.

Africa and Its Heritage: Abu Bakari II

On October 12, 1492, Christopher Columbus and the crew of the three ships under his command completed the *first* crossing of the Atlantic and discovered East Asia or, as later stated, America. And now the horizon of Europe was opened. From side to side. A new line was discovered that foreshadowed others. The situation evolved from a projection that had been fantasized about two months earlier—the expedition set out on August 3—an in-between location whose referents remained unclear: Columbus's Asia. That same year, in early January, the very Catholic rulers of Spain had put an end to the *reconquista* of the Iberian Peninsula. In March, the Jews were summoned by royal decree to convert to the new single credo, Catholicism, or take exile. The result was a massive diaspora, particularly toward the Ottoman Empire. Ten years later, the Muslims of Spain suffered the same fate. Obviously, when Columbus sailed for "America," it was not the politically correct time to be reading Arab chronicles from the previous century. Moreover, Columbus himself was not a model of tolerance. In *Le Livre des prophéties*, he announced that one of the stakes in his crossing, as defined together with Innocent VIII (who died accidentally, or fortunately, one week before the sailing of the caravels)[27] was to find the gold necessary to fund a new crusade scheduled for the 1500 Jubilee Year! However, an Arab chronicle was to tremendously interest the Genoese navigator. It spoke of the pilgrimage to Mecca made by an emperor of Mali, whose reign lasted from about 1310 to 1332. This chronicle was the work of the historian al-Umarī, from Damascus and present in Cairo when Emperor Kankou Musa stopped there, around 1324. In his chronicle, which dates back to the 1340s,[28] al-Umarī speaks about some of the intentions of the emperor, who was a prestigious figure ruling over the vast territory of Mande (Mali), which was a kind of African paradise. Thus, in the *Catalan Atlas* of 1375, the masterpiece of Abraham Cresques, a cartographer whose studio was in the ghetto of Palma de Mallorca, the Mali Empire is listed. One of the illustrations adorning the map represents a black sovereign, seated on a throne planted in the desert. He wields a huge gold nugget toward a Tuareg, who is coming from the west on camel back. Farther west, a ship with all sails unfurled points its bow toward the Rio de Oro, which is located in the present Mauritania. Its crew is Christian. These are the men of Jaime (Jaume in Catalan) Ferrer, a Majorcan like Cresques. The ship and all hands were lost in 1346. The sovereign could only be Kankou Musa, though he became known as Messe Melly for Cresques. The cartographer obviously had sources from the Arab world.

What is the relationship to Christopher Columbus? When in Cairo, there was an inquiry as to his predecessor, and Kankou Musa gave an astonishing explanation. The crown was yielded around 1310 by a predecessor who chose to explore the other side of the great sea. The ambitious undertaking was executed in two stages. The emperor (the *mansa*), anonymous in al-Umarī's chronicle, had first sent scouts off. The scouts disappeared, with the exception of one boat, which returned to bear witness of a strong current that had seized the missing boats. Far from being discouraged, the *mansa* built an impressive fleet of two thousand ships, half of which was for the crew and the other for food. Before embarking on the adventure, he entrusted the scepter to Kankou Musa, who no longer had any news from him. Al-Umarī gave no other details. It is unclear who the *mansa* was, where he had gone, and what links he had with his successor. It would be up to another chronicler, the Tunisian Ibn Khaldun, younger by a generation, to work out the chronology of the emperors of Mali. Then the name of Abubakr appeared, but he had not been specified as the sovereign. The legend of Abu Bakari II, discoverer of the Atlantic, spread in writing in 1912 after the publication of an essay by Maurice Delafosse, *Haut-Sénégal-Niger*.[29] In order to complete the list of Ibn Khaldun, Delafosse said that Abu Bakari was the father of Kankou Musa, without being able to substantiate the claim. We know, however, that African history is not necessarily transmitted through written text. Perpetuated by griots,[30] it is conveyed by oral stories. But the Mandingo tradition provides no place for Abu Bakari II, with few rare exceptions. In *Le Maître de la parole* (1978), Camara Laye (author of *L'Enfant noir*, 1954), transcribed the story of the history of the Mali Empire, which she received in 1963 from the griot Babou Condé, a Belen-Tigui, a Guinean "master of the word." The bulk of the story revolves around the figure of Sundiata Keita, the legendary founder of the Malinke Empire, but at the end of the book, a genealogy of *mansa* is developed. She reported on Abu Bakari II, who is here known as Abu Babari Filânan. He is the one "who perished at sea with hundreds of canoes, which were swallowed up by the blades of the ocean."[31] Obviously, Abu Bakari II is, in the eyes of the griots, a figure of failure. For others, he would be one of the few who opened up space, an extraordinary character.

Could an African have been the first to land in America, nearly two centuries before Columbus? Was his only mistake not to have gone to confirm his feat with the skeptics on this shore of the Atlantic? Who knows? In recent decades, the episode has obviously not gone unnoticed. For some, particularly anthropologists and Africanists of the declining colonial area, African maritime knowledge was insufficient for such a crossing to simply be imaginable. This point of view is now challenged. Favorable currents could very well have helped with the undertaking without allowing for the return. And then, in 1970, Thor Heyerdahl, made popular by his Pacific crossing aboard the *Kon-Tiki*, crossed the Atlantic from Morocco to Barbados on the *Ra II*, a papyrus boat of Egyptian fabrication. Note, however, that between the initiatives of Heyerdahl and Abu Bakari II, there was a difference in size: one knew what he would find beyond the African horizon, the other did not; one faced the known, the other the unknown. For others, who spoke out after the crossing of the Norwegian navigator and scientist, Abu Bakari II had indeed crossed the wall of the ocean, but he had been preceded by . . . the Egyptians. The most famous supporter of this theory is Ivan van Sertima, professor at Rutgers University in New Jersey. In 1976, Van Sertima published *They Came before Columbus: The African Presence in Ancient America*, a monograph in which he exhibited with enthusiasm the possible impact of African (Egyptian) culture on the Olmec, an ancient civilization of Central America.[32] And, in examining the Olmec statues, some specimens found in Mexico represent African warriors. And to say that the construction of the pyramids was inspired by Egyptian techniques . . . Even more, when considering the diary of Christopher Columbus, who would have seen black men in the Caribbean: the descendants of the *mansa* with sea legs? Men painted black? In fact, on the date of October 13, 1492, Columbus reported that residents of Guanahaní had the same color of skin as the Guanches of the Canaries, who themselves were Berbers. I believe, like many others, that it is impossible to form an opinion given the current state of our knowledge about the topic. Perhaps Abu Bakari II had invested America, long after the Egyptians, shortly after the Normans, long after those whose presence does not excite many people but who had the merit of *having been there for sure*: the First Peoples[33] of the North, the South, the Central, and the islands, those in the United States rightly called the *Native Americans*. Speculation about the feasibility or the coronation of the Malian undertaking, often involved a viewpoint both binary (true/false) and centrist (Euro-or Afro-). This viewpoint feeds a faux nationalist discourse, which seeks to establish a territorial precedence, European or African. The eventual success of Abu Bakari II and his fleet would, in any case, be a beautiful African revenge on a history that others have written for it—*written* being the word.

Before concluding this brief overview, I will add my own grain of salt to the sea of doubt. On the margins of the controversies about the reality of the Malinke journey, there is the lingering question of the credibility of the written source of the story: the chronicle of al-Umarī. If the good faith of the Syrian historian is not necessarily called into question, there can be speculation on the significance of Kankou Musa's words. Did the Emperor of Mali exaggerate his story to dazzle his audience in Cairo? Did the allusion to the navigator come from a tall tale? Could an expedition on the Niger River or the river Gambia have been transformed into a fabulous trip? Yet once again, the answer is *maybe*. It is true that Kankou Musa told his hosts that gold grew like carrots in the sands of Sudan. It was noted that such an undertaking would have normally surpassed the narrow framework of al-Umarī's chronicle to appear elsewhere in other reports. The one chronicler to have visited the site in the Middle Ages, Ibn Battūta, stayed from June 1352 to February 1353 at Māllī—that is to say Niani, the capital of the Malinke Empire—during the reign of Mansa Suleyman, Mansa Musa's brother. All this is explained in the last chapter of *Voyages*, by the Tangier geographer, who also mentions Sundiata Keita. Is it possible that after spending eight months in Niani he hadn't heard of Abu Bakari II's expedition? After all, there are not missions to conquer the horizon every day, or even every year! So we can wonder if it was not the chronicler himself who had invented this story to make of it a parable of the human condition: in defying the ocean, would the anonymous navigator have exceeded the limits of the tolerable, committing an act of *hubris* that the wreck punished? *Maybe*, again and again.

The hypothesis is attractive, not because it resizes the impact of the story of Abu Bakari II and Africa's potential, but because it resembles that of Seneca's Jason and Dante's Ulysses, another sovereign who had relinquished power to quench his thirst for knowledge.

Around 1310, the fleet took to the Malinke Sea. At the same time, off the African coast, the Atlantic was an enigma to all Europeans except perhaps the Portuguese sailors who had kept the secret and deterred intruders by monstrous stories. According to a legend that is contemporary to their disappearance, the Vivaldi brothers didn't sink, but rather they circumnavigated Africa before being captured by Prester John in Ethiopia. It is not excluded that Dante was aware of this episode at the time he was writing the *Commedia*. Perish the thought that a parallel can be drawn between European and African sailors plying for the Atlantic off the coast of Senegal and Gambia and their likely failures, which were almost simultaneous. Genoese navigators would certainly not have inspired an Arab chronicler about the story of an African explorer! However, the link between Ulysses in song 26 of the *Inferno* and Abu Bakari II is more robust, especially if you read the passage al-Umarī dedicated to the *mansa* as a parable, an allegory about the willingness of beings to escape the shackles of their condition. Again, nothing says that Dante was directly inspired by al-Umarī, but the reputation of the *Commedia* was already great in the 1340s, and the interactions between southern Europe, including Italy (via Sicily), and the Muslim world were constant. To state the hypothesis very simply, the character of Abu Bakari II is likely to be interpreted as a Dante-esque African Ulysses. Like the king of Ithaca, the emperor of Mali had defied the norm for unleashing a spatial pulsion. The true place of this deterritorialization legend is found, without doubt, in works of fiction. What matters above all, it seems to me, is questioning the motives of the Malinke *mansa*, finding out what could have pushed him to face the unknown from a beach in Gambia. Some contemporary writers share this view. The first of them is the Malian historian Gaoussou Diawara, who chose a fictional solution to transcribe the legend of Abu Bakari II, first in a play, *Abubakari II*, created in Limoges[34] in September 1992, then in a novel, *Avec 2000 bateaux il partit . . . La Saga du Roi Mande Bori* (2000). The second is Jean-Yves Loude, an ethnologist from Lyon, who published in 1993 *Le Roi d'Afrique et la reine mer*, the story of a journey through the Mande in the footsteps of the emperor. The third is the Africanist Alfred Bosch from Barcelona, author of *L'atlas furtiu* (1998), a historical novel written in Catalan and translated the following year into Spanish under the title *El atlas furtivo*.[35] This book traces the vicissitudes Bosch faced by the Cresques, father and son, the same who conceived the *Atlas catalan*.

In his short play, Gaoussou Diawara imagines Abu Bakari dealing with a storm and, even worse with a rebellion instigated by his wife and a griot who sway before the uncertainty of the ocean. The scenario is very similar to that affected at the end of the first voyage of Christopher Columbus: when facing the unwavering persistence of the commander, doubt arises. To the vapidity of the insurgents, prone to supposedly premonitory nightmares and confused by the vision of an undefined and potentially fatal space, is juxtaposed the openness of the former sovereign, which animates the desire to discover otherness: "I dream of seeing our dear Manding going to meet the other, to see our culture feed itself on other cultures, beyond prejudices, and *a priori*. I dream of a world without borders, seamless, where the ideal of brotherhood is not an empty word."[36] Abu Bakari has become a model of humanism. But nothing happens, and mutiny breaks out. It will be controlled, but the commander, mortally wounded, expires at the very moment when the coasts, which he has had the time to baptize Barbary-Brazil, appear. A respite allows him to name the first site Buré-Bambuk, the name of the two gold-bearing cities in Mali that contributed to financing his undertaking. This site will become Pernambuco, now a federal state whose capital is Recife. Officially, the area was discovered in 1535 by the Portuguese navigator Duarte Coelho. Diawara yields in this case to the Afrocentrist mode, which probes American toponyms to detect any African roots.

In *Le Roi d'Afrique et la reine mer*, Jean-Yves Loude recounts a trip he made in 1993 with his wife and a Senegalese friend through Senegal, Gambia, Mali, and Guinea. The three "investigators" traveled across the ancient empire of Mali in the hopes of finding out from the griots what happened around 1310 at the twelfth parallel north, off the coast of West Africa. A specialist of this region, Loude knows that the writings are nothing in comparison to what the griots could teach him. The difficulty lies in the procedures and mechanisms of the unveiling: in West Africa, silence is golden, as danger lurks for both the one who speaks and for the curious listener. Here is the salutary warning that was once received by Djibril Tamsin Niane, author of the epic *Sounjata ou l'épopée mandingue* (1960), from the mouth of the Guinean griot he interrogated: "Wretched one, do not try to pierce the mystery that the Manding hides from you; do not disturb the spirits in their eternal rest; do not go into the cities to ask about the past, because the spirits never forgive: do not seek to know that which is not to be known."[37] The investigator is subjected to the same blame as the emperor: that of overcoming or having overcome the spirits, and with them, the limits of human knowledge strictly enclosed in a familiar space. Loude complained several times about the silence or the diversionary tactics of which he has was the target. But, a man of experience, he knows that this parsimony and these tricks are legitimate because "it is no longer a question for Africa to be stripped of everything."[38] Loude and his traveling companions, however, kept butting up against a wall of silence. The failure is all the more bitter because there is new hope at every step, but it vanishes as soon as it appears. The reasons for this silence are twofold. On the one hand, traditional griots balk at communicating sacred knowledge to the uninitiated. On the other hand, the story of Abu Bakari II is not one that is readily evoked. If the marine emperor readily endorses the hero status in the eyes of a European, he remains primarily a renegade for the Mande. The sovereign turned his back on his responsibilities in order to embark on an adventure sanctioned by failure, a point of no return. For the griots, whether he managed to cross the ocean or not does not change anything. Consulted by Loude, the historian Bakari Sidibe Gambia has confirmed this view: "The griots hardly mention disasters. They do not speak about them, even if they know something. They keep the secret. This is the reason for forgetting."[39]

In *El atlas furtivo*, Alfred Bosch agrees with the griots completely, but he does so by introducing Jaume Ferrer into a story that he edits very little. Abu Bakari is completely ignored. Then we jump directly to the reign of Mansa Musa. The emperor goes to Mecca, making a stop in Cairo. On the way, he spends a fortune and eventually depletes his resources. During his absence, one of his wives, Analia, staying behind in Timbuktu, hosts some pale, exhausted, and ragged foreigners. Their leader, who calls himself Aljauma Fari (Jaume Ferrer, in Arabic), had to go to Rio de Oro. The young woman falls in love with the navigator, and it does not take long for her to get pregnant by him. She is therefore forced to flee with her lover. When they reach the seashore, the couple is imprisoned by the Mossi when baby Selima is born. The captives are delivered to the emperor after his return. An Andalusian counselor recommends to Mansa that he should spare the life of Aljauma Fari and send him beyond the horizon with two hundred canoes manned by skilled Toucouleur sailors. Only one boat returns to announce that Aljauma Fari and his crew had crossed large expanses of water and landed beyond the horizon. Mansa Musa decides to launch an expedition of two thousand boats himself. He disappears into thin air, while Analia, abducted with her daughter by Moroccans, is taken to one of the Fortune Islands, where they live with the Guanches. The story ends with two possible conclusions. One is Malinke: "By vanity, Mansa would not stay with his own . . . By vanity, he plunged into the sea of darkness, like Aljauma Ferrer and as so many other offspring of this earth."[40] The other is Spanish. The narrator is Selima, who, after being torn from her Canary Island is sold on the slave market in Majorca. It is she who tells the story of her family to Jafudá Cresques who, with his father Abraham, is preparing the *Atlas Catalan*. As what is reviled on one side is often honored on the other, Jaume Ferrer and Mansa Musa are immortalized by Jafudá. Thanks to him, the emperor of Mali can hold on to his gold nugget forever.

Today, Abu Bakari II is not exclusively associated with debacle. The influence of the theories of Van Sertima was felt even among the griots. Thus Bala Diabaté, in Guinea, told Loude that Abu Bakari II had reached the American coast, while his predecessor had failed. These are two different sovereigns who have led two different expeditions! And when Loude asked Bala Diabaté what enabled him to affirm that the navigator arrived safely, the griot referred to the indispensable Mexican statues. In Guinea, Loude learned that Abu Bakari II would not have launched his second expedition *despite* the sinking of the first, but rather to go rescue the latter, which washes away the suspicion of presumption, somehow humanizing him. By the same token, he would be deprived of his most original motivation: the Atlantic crossing. Obviously, something is at work in the singular universe of the griots. Abu Bakari II seems to be reabsorbed into the sacred bosom after centuries of exclusion. Why? An early explanation was put forward by a historian of Dakar, at the beginning of the three travelers' stay: "Pathé Diagne painted him as an enlightened monarch, an intellectual prince, attentive to Arab readers of the world: geographers, cartographers, astrologers, travelers."[41] That fed Loude's speculations: "But what purpose is strong enough to push such a continental lord to rise from the throne and to escape on the fluid immensity which is so contrary to his culture? A sincere pulsion towards pure knowledge, or a disproportionate pride directed towards an impossible conquest?"[42] It is this questioning that Gaoussou Diawara confronted in his book published in 2000. Whereas his play portrayed Abu Bakari II as dealing with human and aquatic elements, the novel also offers the genesis of the adventure. It examines the factors that have had an impact on the fate of the navigator. Among them, he identifies a tolerant education, including advocating listening to Arab culture. Lessons coming from the North complement the education imparted by his father, whose stated goal is to make his son a "poacher of knowledge, an explorer of understanding."[43] Far from being isolated, the Mande developed relations with Muslim populations. It goes without saying that Diawara's insistence has a dual purpose: to explain not only the technical mastery of the Malinke ruler but also his concern for cultural openness. It is therefore as an expert, and not from presumption, that the *mansa* launched his expedition. The preparations lasted seven years. In 1312, Abu Bakari II entrusted the reign to his younger brother, who retorted, "But brother, you abandon power for a dangerous and dishonorable adventure. The Griots will say that you fled the difficulties of command. The chroniclers will testify that you resigned."[44]

Diawara makes of Abu Bakari II a champion of knowledge. It goes without saying that the details of the Malinke *mansa*'s initiative are difficult to establish. Loude abstains, Diawara ventures further. But the key is not in the range of facts that have become unfathomable. Abu Bakari is the leader of a long-term project. For Diawara, he represents a model of humanism, "a figure of a prophet in the storm, which, driven by an ideal, chose integrity regardless of unpopularity."[45] But the kind of conspiracy of silence of which the navigator is victim is not just the doing of the Africans. Loude beautifully summarizes the situation: "The world is troubling, reassured Mr. Lion [the nickname given to the author], in the South, we do not speak of this character, and in the North, we refuse to hear about him."[46] And if we refuse to hear about him it is by "this Western intellectual reflex that immediately ejected any bold undertaking born from African initiative, and that under the influence of a simple presumption."[47] Cheikh Anta Diop, a big, bold Senegalese historian, had already reported this clash between intellectual communities, while giving credence to the story of Abu Bakari and the assumption of a first discovery of America by the Egyptians.[48] What would have happened if Abu Bakari II had invested the coast of the Americas? What would have happened if, assuming he had succeeded, Africa had reformed its habits and yielded to the spirit of conquest that has always, and especially since 1492, characterized its small European neighbor? Maybe our view of the world would not be the heir of that which Mercator has crystallized in his *Atlas*. Imaginary spaces are not always there where one believes them to be; sometimes they overlap spaces that conform in all innocence, or by obstinacy, a reality called "objective." Below the twelfth parallel north, in the fourteenth century, the Atlantic was still an imaginary space for Europeans. As for America—maybe neither one nor the other was for the occupants of the two thousand boats of Abu Bakari II. In 1492, Africa had certainly forgotten America, while Europe discovered it . . . or rediscovered it. Reality and fiction, objectivity and imagination are relative. As relative as points of view.

China and Its Heritage: Zheng He

It would be a shame not to mention here, albeit briefly, the possible presence of another man in the "Ocean Sea" during the first quarter of the fifteenth century: the Chinese admiral Zheng He. Born under the name of Ma He in the Muslim province of Yunnan and in a family that made the pilgrimage to Mecca (his father had the title of haji), Zheng He was taken to the imperial capital from a very young age. There he was to live as a eunuch under the name Ma Sanbao. Having become a confidant of Emperor Zhu Di (Yongle), the future admiral adopted a third name, Zheng He, under which he would go down in history. In 1405, he was responsible for leading a first maritime expedition to *Xiyang*, or all the places located west of Malacca, in Malaysia, and in particular on the western sea. In all, he led seven long-term expeditions until 1433.[49] Zheng He, who had certainly reached the coast of East Africa, between Malindi (in today's Kenya) and Mozambique, also landed in the Maldives and Indonesia, in nearly thirty other countries according to the count at the time. His travels were the subject of several reports. In 1597, they even inspired a literary work, *L'Expédition de Sanbao vers la mer de l'ouest*,[50] a historical fiction by Luo Maodeng. According to the British Gavin Menzies, a former submarine commander and historian in his spare time, Zheng He accomplished even more sensational exploits during his sixth trip. He rounded the Cape of Good Hope, which was not yet called so, at the request of Emperor Yongle, who was eager to make known the power of the Ming dynasty (even from a distance) and to raise some additional taxes. Zheng He then sent part of his fleet to the Caribbean, and the others were sent all along the Pacific coast of the future America. In 1423, he returned to China with the remaining ships, some of which had up to nine masts. He envisioned himself receiving a hero's welcome, but according to Menzies, he was disillusioned because the imperial policy had changed: the Middle Kingdom was folding up on itself. The following year, Zhu Gaozhi (Hongxi), who succeeded his father, hastened to put an end to the traveling of large junks. "To prevent all trade and contact with the outside world," writes Menzies, "a strip of land nearly 1130 km long and 50 km wide was completely burned and devastated along the south coast, and its population forcibly displaced towards the interior. Not only were shipyards decommissioned, but the manufacturing plans for treasure hunting vessels and Zheng He's travelogues were also destroyed."[51] These ravages explain why there is no document certifying the authenticity of the admiral's planetary odyssey. Menzies fails to mention that the reign of Hongxi lasted little more than a year. His son Zhu Zhanji (Xuande), who occupied the throne from 1425 to 1435, for his part, financed the seventh and last great expedition of Zheng He, the most ambitious, it seems, though it was not necessarily as ambitious as Menzies says. The latter is based on a series of deductions, cross mapping, and "clues" pieced together to try to establish the route the two Chinese fleets would have taken. It draws on a 1424 portolano chart attributed to Zuane Pizzigano, a descendant of a family of Venetian cartographers, on which appear four islands colored red and blue and located beyond the Azores: Antilia (already mentioned in the l'*Atlas catalan* by Cresques and promising a bright future), Satanazes, Saya, and Ymana. Menzies argues that the first two correspond respectively to Puerto Rico and Guadeloupe. Zheng He spotted these islands before everyone else. According to Menzies, Pizzigano invented nothing. He would have been informed of the existence of these islands by his compatriot Niccolò de' Conti, a trader converted to Islam and a big traveler, who, after passing through Babylon, came across the sailors of Zheng He's fleet in Calicut (today Koshikode) on the Malabar coast, in southwest India. Any such communication between the two, nearly unverifiable, would have been fascinating. In *Beyond the Sea* (2001), translated into French under the title *Ulysse et Magellan*, Mauricio Obregon, who teaches the history of discoveries at the University of the Andes in Bogotá, has reviewed some great sea voyages. He mentions among others the offshore voyages of the Polynesians, such as those popularized by Thor Heyerdahl, but also those that the Vikings undertook toward Vinland. Obregón notes that if the Icelandic and Swedish Vikings had maintained closer relations, they could have informed the Arabs, during one of their trips to the shores of the Black Sea, about the existence beyond the ocean (or of the Ocean Sea) of lands discovered by their Icelandic cousins. So Al-Idrissi, always well informed, would have had the necessary material to add a fourth continent to the map in the middle of the twelfth century![52]

In any event, it seems unlikely that Zheng He and his family had passed through the Strait of Magellan (to some, and the Arctic Ocean for others) before landfall, in California, in Iceland, in Cape Verde. However, after the last voyage of the admiral, the maritime horizon of China was blocked for a long time. I'll confess that I have resisted the temptation to title the paragraph on Abu Bakari's journey "The African Who Discovered America before the Chinese." I'll add that sometimes it was believed that Zheng He was the model for Sinbad the Sailor of whom he was a paronym (Sinbad versus Sanbao, the first name of Zheng He, which the writer Luo Maodeng had remembered). In any case, as Sindbad, Sanbao / Zheng He made seven trips, opened many horizons, and experienced more than once the thrill of space.

America, or the Trinity of Four

Most of these examples place the adventurer who opens space and feels the resulting thrill in a position of power vis-à-vis the limits of the human condition. The hero will fall prey to suspicion and sometimes to charges of outrageousness. The latter is known by *hubris* for the Greeks. For Christians, as clearly shown by Dante, it is pride (*orgueil* in French, *orgoglio* to the Florentine), a word whose apparition is late coming, as it is derived from Frankish. It has an equivalent in Malinke. And there is another one in Chinese: a dictionary tells me that in this language, arrogance is *àomàn*. Zheng He escaped this accusation, but upon his return, he had to admit that it was time to fold in on China, and not to unfold the map of the universe. In the Western world, the context changes as vessels increase incursions *por mares nunca de antes navegados*, to use the third verse of *Lusiades* (1572) by Luis de Camões, cantor of Vasco de Gama's exploits. In Dante's time, the official end of the world was located at the height of the Gibraltar Strait, "To the strait pass, where Hercules ordain'd / The bound'ries not to be o'erstepp'd by man."[53] *Non plus ultra* was the inscription that the ancients believed to be engraved on the columns bounding the Mediterranean and the space to the West, between Calpe and Abila (for us, Gibraltar and Ceuta). Dante had merely transposed it into Tuscan to better castigate Ulysses's thirst for knowledge. Dante would not have approved of the diver in the fresco at Paestum.

In the sixteenth century, the inscription was changed by Charles V, whose motto became *Plus ultra*, "Further Beyond." It is true that the Spanish peninsula, in the southwest Atlantic, already stretched "further beyond." Columbus had sailed from a port beyond the Pillars of Hercules. America was even "further beyond" on the other side of the same ocean. It was still necessary to ensure that the ocean was not a river. Everything is definitely relative. In *Historia general y natural de las Indias*, the first part of which appeared in 1535 (the following two parts were not published until the nineteenth century), Gonzalo Fernández de Oviedo already considered that the Mexican coasts to the south of the Americas were a modern *Mare nostrum*. New Pillars of Hercules were designated: they now stood in the Magellan Strait. In turn, they were soon to be bypassed. Embroidered in gold letters on two ribbons encircling stylized columns, the two words of the emperor now figure on the Spanish coat of arms. In 2008, a European military force under French command went to the Chadian Savannah. The group was baptized *plus oultre*, further beyond. Some sayings are decidedly tough-skinned. Did the French army want to make a discreet tribute to the empire on which the sun never sets? Charles V would have liked that. He spoke in French, by the way. What proof do we have? It is he who supplied Hernani's reply in Hugo's play.

It is not Hugo but rather Rabelais of *Gargantua* fame about whom we should think here. In chapter 8, after a long discursive excursus intended to ridicule the work of heraldry, the Rabelaisian narrator decides to interrupt. He uses these terms: "But my little skiff alongst these unpleasant gulfs and shoals will sail no further [plus oultre], therefore must I return to the port from whence I came."[54] Following this example, let's return to our starting point: infinity. Moreover, the path that leads there is short: this infinity could find itself in the work of Rabelais and, more specifically, in the mouth of Pantagruel. At the height of the war against the Almyrodes, a downpour surprises the giant's army. To protect his troops from the storm, Pantagruel halfway sticks out his tongue. All the men find refuge under the deployed organ, except Alcofribas, the narrator, Rabelais's alter ego, who decides to enter the mouth of the giant. He walks on the tongue, sees big rocks, large meadows, large forests, "strong and large cities, not any smaller than Lyon or Poitiers."[55] In short, a new world opens before him, of which hitherto he had not even suspected the existence. And this world hides another one. The narrator, "all flabbergasted," starts a conversation with a peasant who is in the middle of planting cabbage. Then he asks if there is "here a new world," and the man replies: "Sure, said he, it is never a jot new, but it is commonly reported that, without this, there is an earth, whereof the inhabitants enjoy the light of a sun and a moon, and that it is full of and replenished with very good commodities; but yet this is more ancient than that."⁵⁶ What is this other world? This is of course the world that appears outside of Pantagruel's mouth, the world of Alcofribas/Rabelais, which, for the planter of cabbage, comes from discourse of the legend. We could imagine that this is yet another world, neither that of the narrator, nor that of the peasant, that revolves around the city of Aspharage, a world that is a *mise en abyme* in another part of the giant's body: his stomach, for example, from which emanate pestilential exhalations plaguing the cities of Laryngues and Pharyngues, which one does not make the effort to locate. The cascading of worlds becomes dizzying. One thing is sure: in chapter 32 of *Pantagruel*, the great humanist had formulated the beginnings of a theory of possible worlds. Compared to the theorists of the twentieth and twenty-first centuries, however, there remained a difference in size (which I do not link to the giant). The possible world of Rabelais was not just an abstract construction, purely verbal, and self-referential. He had a tangible referent in the new lands discovered. They had a floating status: they inspired a description that aimed to be "realistic," but they only began to exist themselves in the story made for the listeners and readers, who were more and more numerous, of Old Europe.

Without being quite able, I promised myself I would resist the temptation to talk about the "world enclosed by the mouth of Pantagruel" and for a very simple reason: Erich Auerbach has already masterfully done so in *Mimesis*. Here is an excerpt of the commentary he made about the conversation between Alcofribas and the peasant: "The man speaks of the 'new world' as the people of Touraine, or any other region of Western or Central Europe, probably spoke about recently discovered countries, such as America or India."⁵⁷ With a little imagination, one could portray Touraine as the New World of the peasant. Where is the gold? Who should be converted? What should be thought of Amboise, Chenonceau, and the other outposts of a French Eldorado? That would make a good subject for a novel. Well after Rabelais, Fernando Iwasaki, a Peruvian writer leading a flamenco school in Seville, came to a variation on this paradoxical theme in a book called *El descubrimiento de España* (1996). But, as for the author of *Pantagruel*, the discovery of the New World was mostly an opportunity for Iwasaki to speak better about the Old World, which here would be Peru, in a logic consistent with that of Rabelais. And as contemporary literature is increasingly fond of these reversals of Western perspective, outlined by Rabelais and later developed by Montesquieu in *Lettres persanes*, I will not deprive myself of mentioning another South American writer: Federico Andahazi. In *El conquistador. Un azteca descubre Europa antes del viaje de Colón* (2007), this Argentinian author imagines that a young Aztec, Quetza, sets sail with some men to check his intuition that there is land on the other side of the Tenochtitlan coast.

It seems unnecessary to me to insist on the profound upheaval that the Renaissance brought about in the perception of the world. Just about everything has already been written on this subject. In fact, it is necessary to examine an element that I have neglected: the progress made in reading the sky and the stars. The world could no longer be confined to places that were known. "It is indeed," writes Augustin Berque, "due to the Copernican revolution that the universal stretched out to oust the worldly; space replaces country, while all the while deconcretizing the world; it all flowed logically from the fact that the infinity of the space of the new cosmology radically suppressed the home and the horizon which were the founding benchmarks of the ancient world."[58] With the reflection of a heavenly universe expanding, which astronomers didn't cease to map, the earth was bound to be larger than the gaze and reason could embrace. New seas were roamed; new islands, new continents were discovered. Others yet could still be found somewhere else, out of reach, but not for long. The appetite was immense. Literature would attest to such. Utopia was born to translate into writing the content of an in-between: it narrated the state of an unknown space that transformed into a familiar place to men. *Utopia* was not an imaginary space, in the traditional sense of the term: it was a space called to "create" itself one day. In sum, the fictional representation preceded sometimes, even often, the geographic referent. The discovery confirmed *a posteriori* the imaginary projection. The most spectacular case is undoubtedly that of Australia. We had conjectured the existence of a *Terra Australis Incognita* since ancient times. Much later, in 1676, Gabriel de Foigny wrote a utopia called *La terre Australe connue* in which the word *Australia* appeared. In the early nineteenth century, the British began more often to couple the toponym *utopian* with the southern continent. In 1824, it officially became *Australia*. And the loop was closed. At the end of an age-long process, Australia embodied a utopia that asked only to find its geographic fulfillment. The gap between fiction and reality is far more tenuous than the positivist approach would suggest.

There is a shift from a closed world to a virtually infinite world. The *Terra Australis Incognita* embodied for some time—over the span of a few centuries!—the possibility of a fourth continent whose antipodal presence would have balanced the anchorage of the three others in an unstable universe. This fourth continent, quickly imagined, was the last to be discovered. Meanwhile, another fourth continent had emerged from the waves, but it was discovered before it was conceived. Even if we set aside Abu Bakari II, Zheng He, and all the Vikings in the world, we can still ask if the continent was really discovered by Columbus. America? The Genoese thought he had discovered the eastern part of Asia. When he touched land on the other side of the Atlantic, he thought he was treading on the floor of an archipelago situated off of *Cipango*, Japan. In October 1492, the error was human. But, for Columbus, it quickly became evil. Right to the end, with a blind determination, the navigator continued to believe that he had opened the way to Asia by the west. Cuba passed for the Malay Peninsula and the other Caribbean islands imitated, for him, the geography of the Asian archipelagos. In his evocatively titled essay, *La invención de América* (1958), Edmundo O'Gorman drew the following portrait of Columbus: "He arrived in Asia; he was in Asia and he came back from Asia. Nothing and no one would succeed in changing his mind until the day of his death."[59] During his third voyage, in 1498, Columbus discovered the coast of Venezuela and with it a tract of land decidedly incompatible with the configuration of Asia, even if freely interpreted. As explained by O'Gorman, the navigator was faced with an unsolvable enigma: this area was exempt from any known reference; no authority had mentioned it. He didn't know what to do with it. So he made it into the headquarters of heaven on earth. The novelty of the place was therefore an illusion; it had once housed the germinal cell of human society. In any case, if the north of "America" could correspond to southern Asia, the south of the "New World" was a blank space. The quotation marks concerning America and the New World are de rigueur, but one thing is established: Christopher Columbus was the first to enter into a completely blank space, of which no literature, no authority had announced its existence. He didn't do it in 1492. There, he discovered a road rather than a continent, because, after all, we already knew . . . Asia. The evidence of the novelty appeared to him in 1498. But he could not bear the impact, and the uncertainty even less. Like Dante's Ulysses, he had extracted from the new world a paradise on earth and then sunk. Oh, certainly not physically, but morally: he returned from his third trip under the charge brought against him by the emissary dispatched by the king to Hispaniola (Santo Domingo).

The voyages of Columbus proffer several lessons. First, they demonstrate a return to an old worldview where the displacement of limits took place in the west, according to the course of the sun. Among the Greeks, one projected *plus ultra* toward the River Okeanos: it is the process *exokeanismos*, which I mentioned earlier. This is where the spaces were blank. In principle, they were even outside the reach of human control; they were the realms of the gods and of the nymphs. To the east, where the sun rose, the day grew old very quickly, too quickly even. It could not conceal important novelties. It had the wisdom of age; it brought a confirmation and a form of assurance. The Genoese navigator had nevertheless complicated this vision: he sought the rising to the West. He did not, strictly speaking, have a "Western" vision. Yet again, the originality of Columbus lies in his choice of an alternative, even paradoxical, route to the Orient. The aim of his visit was not the discovery of a space but a different quest of known locations: fabulous India, Marco Polo's and Kublai Khan's China. That's why he never gave up on making "America" into a miniature East.

His involuntary discovery of a new continent produced an effect that he could not conceive of and of which the symbolism of numbers explains. As noted by O'Gorman, the three are abandoned in favor of the four. The addition of a new continent involved obviously not just a simple addition of an abstract arithmetic. The three referred back to the Trinity, the Magi, and the sons of Noah (Shem, Japheth, Ham), whose names were associated with the three "historical" continents and their inhabitants. As for the rest, medieval geography relied on a doctrine. The four, however, did not evoke anything, except perhaps the cardinal points. It escaped the biblical dynamic; there was some indication that man had left his "old cosmic prison"[60] to affirm "his sovereignty over the universal reality."[61] The question, for Columbus as for his successors, would be to know what relationship the people of the fourth continent were to have with the God of the Bible. It is in any case clear that the men and women of the three "old" continents no longer held the monopoly on humanity. Therefore, everything became possible. The perspective of infinity appeared, which confirmed on another plane the increasingly insistent gazes fixed on the stars, the naked eye being replaced in the early seventeenth century by the first telescopes.

Between 1511 and 1520, Carpaccio painted a series of five paintings (one of which has disappeared) on the life of St. Etienne. They were intended for the Scuola di Santo Stefano, in Venice. Around 1514, he produced *La Prédication de saint Etienne à Jérusalem*, which is held today at the Staatsgalerie in Stuttgart. In the foreground, we see the saint preaching in the temple of Solomon, standing on a pedestal in ruins, pointing to the sky. Marble debris litters the ground: paganism is defeated, it would seem. An audience composed of twenty very different figures, whose attire recalls (with complete chronological indifference) the three great monotheistic faiths, listens to him, and some comment to each other. In the background an "Italian" landscape unfolds. Rolling hills are covered with regularly shaped buildings, which could all be Tuscan if they weren't topped with crescent moons. This harmony contrasts with the disorder in the foreground. Men wearing turbans and chatting quietly, standing still or walking, occupy the intermediate space of the painting. So much peace and harmony are surprising. What more could the Christian God bring? What perspective could the preaching of the saint open? In fact, the Acts of the Apostles provides an answer: "But he, being full of the Holy Ghost, looked up stedfastly into heaven . . . Behold, I see the heavens opened, and the Son of man standing on the right hand of God. Then they cried out with a loud voice, and stopped their ears, and ran upon him with one accord, and cast *him* out of the city, and stoned *him*."[62] There where the viewer of the painting sees a worldly landscape, Etienne sees the sky and the infinite *ultra*worldly. Did the stoning of the holy man by the public follow the awareness of "collateral damages" that was engendered by the Christian vision of heaven? Certainly, this malicious interpretation did not cross the mind of the Venetian painter! For Carpaccio, Etienne had abolished the limits of the world. He offered immensity in exchange for the harmony of a peaceful yet closed, and therefore imperfect, place. His public conceived terror, even panic. They killed him, out of town, in a place without hearth or home.

A suspicion remains, however, a residue of ambiguity. That the sky would open to Etienne's listeners and reveal the infinite was without doubt a fact—although only some of them, three or four, including a woman sitting in the geometric center of the group, looked up. But what were they supposed to find? God? When one considers the serenity that characterizes the small group revolving around Etienne's pedestal, we could say that God's presence was not really an ontological necessity. The message that the saint conveys seems somewhat out of cadence with his environment. He himself seems stuck on his pedestal, offset to the left of the painting. As noted by Michel Serres, in *Esthétiques sur Carpaccio* (1975), his face is subdued compared to that of the ephebe (beautiful young man) that is placed next to the pedestal. Etienne is like a statue. He became the image of something old, perhaps something even obsolete. There is the strange impression that the man is going about his business regardless of Etienne's configuration of heaven. Is the infinity that manifested itself before the preacher void of any divinity? Would it now be promised to man and man alone? The celestial infinity, which is also a reflection of the infinite sea—would it be a human space to explore? This revelation caused panic. It is this revelation that the crew of Christopher Columbus's three ships must have experienced when they were beset by doubt in the middle of the Sea of Darkness. Was Columbus, like St. Etienne, with his finger pointing to infinity, just as represented by the statue erected in the center of the *Plaza de Colón*, in Barcelona? The enthusiasm comes later, when the whole point of crossing the boundaries of a closed world appears. Perhaps it is this third reading—so implicated with consequences—that would cost Etienne his life, the first martyr of hagiographies. In the *rinascimentale* version of Carpaccio, Etienne had perhaps understood that *sense* had become a polysemous concept. In his stimulating analysis of the painting, Serres noted that "the meaning is at least two-fold: the spatial and the predicated. The acute vector and the nodal place designated by the index of the preacher." And he goes on about Etienne: "But still, what does he see? Heaven. Knowledge, even space. A division of space. But really, what does he see? I see the heavens opened. Namely an open space."[63] For Serres, space and the predicated are one. As for me, I'm not sure. In Acts, the meaning was single and unique: that which was preached. The emergence of a spatial sense was peculiar to the new era. It expressed the infinite terms of open space. It gave a new importance to cosmography. With a world called to return to humanity, as Frank Lestringant said, to "return to man this whole part which is the earth, that is the holy task which employs cosmographic labor."[64] Etienne was the holy herald of God; the new hero, who is only the herald of himself and of the political power in the service of which he is attached, is no longer confined to a *one-way* communication with the deity. He proclaims his own truth. His "holy task" will be to reconcile the universe described in the Bible (which was still an authority) with the new maps of the universe, terrestrial and celestial, of which he was trying to sanction the gap.

The World's New Clothes

The lexicographic evolution provides us yet again valuable benchmarks to understand better the *forma mentis* of the time. I turned to Paul Zumthor, who executed a philological study of the "trip" in *La Mesure du monde*. In the Middle Ages, the traveler was not the one we know today, and the journey wasn't either. The trip was a well-marked itinerarium, a progression of stops. In the modern sense, the journey only entered common vocabulary at the end of the fifteenth century. Within half a century, Du Bellay's Ulysses would not have "a nice trip." He would have continued to founder, as Dante's Ulysses. In English, the prospect of the trip could hardly be more encouraging. Travel comes from *travail* (work), but work implies torture, painful childbirth, or at least a strong constraint: the *tripalium* was a yoke under which oxen were coupled, among the Romans. In German, *Reise* comes from the Germanic *risân*, which means "elk," but also "break" (found in the form *Riß* in modern German). Throughout the Middle Ages, the journey/travel/*Reise* is a "passage torn from here," in the words of Zumthor. Sometimes you have to tear yourself away to move from *here* and go *there*. In French, the word changed meaning and connotation at once when Columbus's caravels sailed to the New World. The route/path is characteristic of a geographic view based on a rational accumulation of places. Traveling itself is associated with an entry into space. The *homo viator* is a *spatiator*. As Zumthor writes elsewhere, the *homo viator* is not an absence but a distance. And distance is something you can overcome. But let's not stop in the middle of our path. What about "geography"? The concept is not new. The Greeks spoke of "geographers" since Eratosthenes,[65] and Strabo was the author of *Geôgraphiká*. But it took some time for the dictionary to integrate these new words. *Geography* was coined in its humanist form in 1513. It was thus inaugurated only a year before Etienne raised his eyes to the infinity of space. And Zumthor says, a priori without reference to the painting by Carpaccio, "It is only then that it aspires to become a science of terrestrial space, at the same time a desire to control space takes shape and power."[66] I will speak shortly about this desire to control. Zumthor says, "In the 16th century, the reference to the Ancients is merely a style. At the same time the authority of the Bible is weakening in this domain. Geography is, for many centuries, an experimental science of space."[67]

The resurgent world left the convenient shelter offered by biblical geography. In the Middle Ages, as formulated by Zumthor, "Moses is a geographer *par excellence.*"[68] From the late fifteenth century, Moses ceases to be a geographer. The world has become too vast to be embraced by a glance; it was thrown down from the top of the 2,200-plus meters of Mount Sinai. Paradoxically, although this paradox is apparent, we establish the notion of the horizon when we are convinced of the extreme amount of space that extends beyond the line at the end of the gaze. According to the little arithmetic that I hold so dear, Moses must have seen 170 kilometers away in the distance—or maybe a little more, if we consider the erosion of Mount Sinai. Much more, if one believes that the view he enjoyed was inspired by his God. It is the optical illusion from which only prophets know how to benefit. Their geography is not that of ordinary mortals. However, at the end of the Middle Ages, they began to boil over with impatience. It was their responsibility to establish themselves as masters in a world that remained to be traced out.

Moses was the geographer of the Judeo-Christian civilization of the Middle Ages; for the Greeks, the great geographer was undoubtedly Io, priestess dedicated to the worship of Hera. Her story is well known to readers of Aeschylus (*Prometheus Bound*, *The Suppliants*) and some other Greek and Latin authors, including Ovid and Virgil. An unrepentant seducer, Zeus started off one day to make her his own, provoking the jealousy of Hera. To help Io avoid the vindictiveness of his wife, the god transformed the nymph into a white cow. Hera was not to be fooled. She placed the transformed Io into a fenced enclosure under the supervision of Argos Panoptes, a creature with a hundred eyes. At the request of Zeus, Hermes managed to free Io after killing the jailer, whose eyes now decorate the peacock's tail. But once again, Hera did not allow herself to be had. She sent a horsefly to the heels of the cow, who, becoming crazy, began a frantic race through a sparsely staked space. She went to Dodona, and from there, on to the shores of the sea that bears her name since then: the Ionian Sea. She then swerved toward Caucasus, where she met Prometheus tied to his rock at the end of the world. Then she crossed the Symplegades Strait, whose stabilization in a more human geography would be the work of the Argonauts. However, she also gave it her name: it became the Bosphore, which "carries the cows." Io continued her stampede to Media, in India. She then deported to Arabia, Ethiopia before arriving in Egypt, where Zeus gave her a human figure. There she established the cult of Isis. She had Zeus's son, Epaphos, who would be the ancestor of Agenor, himself the father of Europe. Europe was the princess that Zeus, transformed for the occasion into a white bull, would later ravish on the coast of Sidon, in Phoenicia, to take her to Crete. The story of the abduction of the Phoenician princess gave birth to the following great geographical myth, the one to which belongs the explanation of the origin of Carthage, Thebes, and Cilicia. Europe is a distant relative of Io. The plasticity of the Greek myth is always admirable. Io's attitude is diametrically opposed to that of the Renaissance explorers. Io has been subjected to an itinerary whose extraordinary nature stands up to the challenge sought: make Greece, against all evidence, the source of inspiration for the Egyptian pantheon, which is much older. Io created Isis after making a world tour; in a different sense, she made Epaphos who is linked to the god Apis, who had the form of a bull. It is certainly necessary to identify all the bulls, cows, and oxen whose fates have had an impact on ancient geography, from Greece to Carthage and Rome. In any case, Greece indentures the Egyptian precursor, whose link to the story of Moses is known.[69] Io, choosing nothing, fled; Jason chose to leave to meet the challenge of a tyrant; Christopher Columbus and his successors have chosen for their part to take the tour that opened a new world. There was no question of making history, or colonizing the story of a famous predecessor. Christianity was sufficiently assured of its aura and authority, especially in Spain in 1492, after the *reconquista*. Rather, it was to expand the world in order to universalize a type of civilization that was thought to be unique and intended to serve as an undisputed model. That, in fact, this model was questionable only dawned on a few enlightened observers later, when we began to assess the extent of the suffering of the "Indians." We refer in this regard to the pioneer writings of Bartolomé de Las Casas and especially his *Brevísima relación de la destrucción de las Indias* (1552), written in Valencia.

During the historical watershed brought about by the transition from the Middle Ages to the Renaissance, the surplus world that opened before the eyes of sailors appeared as a bare space. That seemed immodest. It was no longer a time when a naiad, even transformed into a heifer, could travel the world in the nude. It fell to her contemporaries to find new clothes. I have already established the link between the map and fabric. Let's return to it for a short time. The medieval word *mappemonde* comes from the Latin *mappa mundi*, in which *mappa* refers to a napkin. In Horace, it is a tablecloth; for Quintilian, it is a piece of cloth that is thrown into the ring to signify the beginning of the games. If we placed this etymology on the metaphorical scale, we could summarize the issues quite well at the beginning of the era of conquests: it is perhaps by throwing a blank map that needed to be filled out into the arena of the world that we unleashed the cruel game of colonization. In *mappemonde*, it is not just *mappa* that refers to a textile etymon, as among the Romans *mundus* was not only the "world" but also "finery" or "ornament." In that, the Latin merely imitates the Greek: the *cosmos* covered the same two senses, which converge in the idea of order or harmony. Many modern languages have retained this ambivalence, creating first a cosmography and then also a cosmetology, one and the other wanting to be scientific, one and the other making up appearances. The *mundus* and the *cosmos* became necessary when it came to covering space with a type of clothing, an artifact to hide its own nature. We return to the division of Jacques Le Goff between the geography of nostalgia, reserved for the space dreamed of by the imagination in a closed world, and the geography of desire, which faces the open. Pure space is staggering; it is excessively "fascinating"; it enters a geography of desire. Jason must have felt this fascination upon his exit from the Symplegades; Columbus, too, on his approach to the "New World," even if it is not stated in his logbook. Why such modesty? Maybe because in "fascination" there is *fascinus*, which was the penis erected by the Romans. During the Renaissance, it took a *mundus* to hide this *fascinus*.

CHAPTER 4

The Invention of Place

Cabeza de Vaca or the Possibility of Space

There has already been much talk of Christopher Columbus, and I will be no exception to this rule. There is also much talk of Cortés and Pizarro, conquistadores and executioners from Mexico and Peru. By virtue of the cinema and the film made by Werner Herzog in the early seventies in the Peruvian Andes, there is sometimes talk of Lope de Aguirre, the "Wrath of God." However, we almost never talk of Alvar Núñez Cabeza de Vaca. However, Cabeza de Vaca is a great man, even a very great man. He felt the dizziness of space, and there were not so many to do so. Several trails lead to him. We could learnedly cite some American essays, or, more recently, two or three pages of *La Conquête de l'Amérique. La question de l'autre* (1982) by Tzvetan Todorov. Another possibility is offered by a film from the Mexican director Nicolás Echevarría whose title matches the not so banal surname of the *conquistador*. His name, Cabeza de Vaca, or "Cow Head" is a maternal inheritance and not an unusual sobriquet. In 1991, at the height of the wave of often mediocre blockbusters devoted by the Hollywood industry to the "discovery" of America and its fifth centenary, Echevarría proposed a different reading of the event, one that is more respectful of the views and culture of the First Peoples. He attributed the lead to Juan Diego, a great Spanish actor whose interpretation was nearly performance art. It fell to Diego to embody a man confronted with the problem of communication and cohabitation with other men, natives of the current Texas, whom he knew were separated by a radical otherness. At the time, given that not many attended, universities did not teach anthropology or intercultural studies, and the notion of exchange with the Other was related to an anomaly. This is the essence of the extraordinary adventure of Cabeza de Vaca. The usual title of the Spanish text is *Naufrágios*. It was printed in 1542. In French, it was translated in a rather flat way: *Relation de voyage*. But it was translated . . . and that in itself is a lot. Echevarría's film transposed the story in its own way, often brilliant but not consistent with the story given by the character of principal interest. Echevarría was certainly inspired by the atmosphere of Alejandro Jodorowsky's films, the Chilean director who has worked in Mexico and who, in *El topo* (1970) had portrayed the desert with a baroque emphasis and initiatory concern that announced the spirit of *Cabeza de Vaca*. There are alternatives to Hollywood.

Cabeza de Vaca accompanied an expedition to Spain as its treasurer on June 17, 1527, under the command of Panfilo de Narváez, a ferocious man who had lost an eye in a vain attempt to stop Cortés a few years earlier. A total of five ships set sail for Florida, which the conquistadores already knew, but which they had not yet succeeded in seizing. After a stopover in Havana, where they underwent a hurricane, the small fleet landed on the shores of Florida on April 12, 1528. There the carnage began. Out of all the men who landed in Tampa Bay, only four returned—miraculously—to Spain. Narváez gave his ships, still without anchorage, to crews reduced to a minimum in order to follow the interior path with the infantry and forty horsemen. As usual, what was at stake were gold and possessions. Great evil attacked the Spaniards; after six months of wandering in a hostile environment, the losses were already numerous. The conquistador, who knew he was cut off from his ships, resigned himself to build makeshift rafts. And a new drift began for 250 survivors. The rafts skirted the coasts of Alabama and Mississippi, then Louisiana. Thirst and hunger prevailed, and some had resolved to eat each other. In learning about it later, the natives were scandalized. (They are not necessarily cannibals just because they are suspected of being so). In November 1528, after two months of floating around, the boats wrecked, one after the other. That of Cabeza de Vaca sank in the vicinity of Galveston Island. Some men survived before dying or disappearing forever. Four men made it through: Cabeza de Vaca, Alonso del Castillo, Andres Dorantes, and Estebanico, Dorantes's slave and perhaps the first African to set foot in the future United States. This wreck was also the first in a long string of dramas of which Galveston (Mal Hado, "Evil Spell," for the Spaniards), was the theater. Several centuries later, in 1900, a tragedy took place there that deeply affected the minds of the inhabitants: a hurricane devastated the island and caused the deaths of nearly eight thousand people.

Soon, Castillo, Dorantes, and Estebanico were separated from Cabeza de Vaca, who lived alone in what would be for a Spanish man the exact equivalent of nothingness. It was only much later, during the winter of 1533–34, that he found the other survivors by chance. And it was not until July 1536 that the four men came across a Spanish patrol in New Galicia, in northern Mexico. They had spent nearly eight years in the company of natives of both sides of the still virtual border that would one day separate the American Southwest and the Mexican *Noreste*. Their adventure stands out in the annals of conquests.

Fortunately, Cabeza de Vaca's story is in many respects unique. Having just arrived on solid land and been reduced to captivity, the castaway was faced with unlimited space. In Texas, the sensation must have been as intense as the horizon was vast. As Yves Berger wrote in his preface to the French translation, "Cabeza de Vaca, one suspects, had only one idea in mind: to escape from this huge prison made for him by Indian space and return to the White world—his world."[1] Space is likely to be perceived as a prison. The infinite vastness of the Texas prairie meets the infinite smallness of the jail cell (like that tiny one, in Fort Sill, Oklahoma, where Geronimo spent the last 15 years of his long life). This is especially true when this space offers so little to the occupants of Texas at the time: the locals ate on average once every three days and suffered from a persistent cold, because it seems that the climate has changed in the meantime. Their feasts consisted of mixing some berries with dirt and grinding them together in a hole. The idea of digging deeper had not yet occurred, and anyway, there would have been the question of what to do with the black goo springing up from underground. But Cabeza de Vaca did not resign himself to undergo a paradoxical confinement. He conformed to the dimensions and the conditions of the space. He agreed to leave it open. Practically, he became naked. But at the same time as his clothes, he was stripped of some of his prejudices. He became a shaman and introduced on-site trade (the market economy!). He acquired a certain kind of freedom of movement. In this deterritorialized space, amazing things happened, from a European point of view. Endowed with powers that his otherness rendered supernatural, Cabeza de Vaca was able to heal the sick. He even accomplished a semblance of resurrection. Conversely, due to a feigned mood swing, he involuntarily caused the death of eight men. The Spaniard was practicing psychoanalysis without knowing it, as Jourdain explains later. But he was not ridiculous. He faced what was for him a parallel reality. He defied the suspension of bearings, standards. He accepted the challenge, and this allowed him to probe space and pure humanity with it, through it.

Having left as a conquistador, though he was immediately more reasonable than the others, than Panfilo de Narváez in any case, Cabeza de Vaca was able to durably tune into absolute space. He therefore did better than Jason, and much better than Columbus. This enabled him to resist. The end of his journey coincided with the meeting of a patrol, announced by a series of signs, indications of "civilization"—that is, traces of a European presence. What the conquistador and his three companions entrusted to those listening to them merits to be transcribed: "Along the way, we always found more information about Christians, and we told the Indians that we were going to try to find them to tell them not to kill or enslave the Indians, nor to dislodge them from their lands, nor cause them any harm, to which they greatly rejoiced. We passed through large spaces that we found completely depopulated because their inhabitants had fled to the mountains."[2] Did the speakers believe their own word? Somehow, Cabeza de Vaca and his companions in misfortune had become humanists in practice and act. Their compatriots and contemporaries were, at best, humanists in theory.

When the four survivors of the Florida expedition emerged again in the territory controlled by Diego de Alcaraz's Spanish troop (in the present province of Sinaloa), they were not so easy to recognize. Long, suspicious glances came their way. They had to quickly be dressed, although they could no longer stand clothes or mattresses. And the large, depopulated areas that Cabeza de Vaca, Castillo, Dorantes, and Estebanico had filled in their own way fell under the control of the new masters. Nicolás Echevarría's film ends with the symbolic image of a large shimmering cross, transported in the desert under a blazing sun. To conclude the great fable of Cabeza de Vaca, it is my duty to say a few words about the fate of the protagonist. He returned to the Americas after transcribing his first adventure in Spain. He discovered Iguazú Falls, at the current border between Brazil and Argentina, before settling in Asunción, founded shortly before. In Paraguay, he spared the "Indians" a little too much for the settlers' taste. For this he was arrested. Despite his title of governor of Río de la Plata, he was considered a traitor. He was escorted to the motherland in irons. He escaped death but was exiled to Oran before being pardoned and dying in poverty. Yes, this conquistador, who conquered what the Germans would call the *Einsicht*, that insight that can "see inside" of things, had penetrated the true nature of space. Undoubtedly, Cabeza de Vaca was a very great man.

Columbus was used to "baptizing" that which unfolded before his eyes. To him, religious vocabulary was ad hoc. Cabeza de Vaca, for his part, respected the original names and reproduced them to the best of his protoethnographic skills. Speaking of the First Peoples, he wrote: "I want to say their nations and their languages, from the island of Mal Hado until the last."[3] And he kept his word, to the extent of his abilities. Columbus tried to build the place on the spot— and this building continued to feed a religious isotopy. Cabeza de Vaca strove to spare the option of an open space. These two men embodied two different visions of the new world that were not necessarily irreconcilable. That of the Genoese conformed to the desire to transform the new lands into a stable and sedentary territory along the still provisional border (which introduced nevertheless a margin of oscillation in the process). That of Cabeza de Vaca was ready to join a nomad song, which escaped from the norm. His vision apprehended different lines of thought: that of the Caoques, the Doguenes, the Guaycones, and of all the other First Peoples encountered. This vision was plastic; it finally came around to bringing forth the idea of difference. It was ready to take into account an original *modus vivendi*, which assumed that far from having to be presented in one single mode, the Spanish, Christian, and Occidental lives constituted a paradigm, which was, of course, but one among others. Of course, once placed under the orders of Panfilo de Narváez, the conquistador was not intended to identify himself with a nomadic model that would find justification only through the concern of mobility *in se*. In fact, no form of nomadism, to my knowledge, is motivated by a gratuitous desire to move around. Mobilization takes place as part of a material quest and, more rarely, a mental or mystical quest. Cabeza de Vaca was experiencing nomadism, to which he subjected himself, in the sense that Zumthor has identified: the nomad is not one who opposes the idea of the home, but rather one who departs from a "classic" notion of space—that is from the vision of space conveyed by sedentary peoples. For Zumthor, the nomad is one who is "living off of the premises, or rather, for him, place unfurls in space."[4] The nomad goes beyond the traditional divide between place and space; his place is in space, and that space is never abolished. Then what Zumthor called disaffiliation happens, which is a sliding out of the norms. And then the Swiss medievalist refers to Cain, the first nomad, as pointed out in Genesis, but also the constructor of the first artifact place, the city of Enoch. Was Cabeza de Vaca the Cain of the Americas? Nothing is less certain. He did not kill his brother; he even made every effort to save him. His fellow adventurers knew that, as did his hosts.

Cabeza de Vaca could have ushered in a different reading of these *Indias* that were about to be reterritorialized on something known, familiar. His gesture could have introduced an equivalence between the new world and the possibility of a new perception of space. It could support the fact that the gap was bearable, that the lack of reference, dizzying in itself, was compatible with the human condition. It could have shown that the development of life in what Deleuze and Guattari call an "intermezzo"[5] was possible. It would have required that the new world become a laboratory of refound nomadism. The peoples who inhabited southern Texas were mostly sedentary. But through the ligation activity that he operated between clans, and by his constant concern for mobility, Cabeza de Vaca had succeeded in empowering a nomadic dimension. One could certainly retort that the ultimate aim of this move was to return to *civilization, true civilization*, the Spanish one. But could we be certain? For the uprooted conquistador, the *true* part had become a relative value. As Pilate before him, a question was now pestering him: *Quid est veritas*? The description that Cabeza de Vaca has given of his nomadic years contains a form of enthusiasm that is almost shameful. It is discreetly indicated by the hint of bitter disappointment that accompanies the tale of the unexpected reunion with the patrol somewhere in a Mexican desert. Perhaps he had finally found an answer to the question that the procurator of Judea had formulated prior to himself. Pilate was also a conquistador, though, oddly, the Romans did not have a word to express the concept. In the West, it is in fact the medieval man who seems to have realized the nature of the "conquest." The verb *conquérir* (to conquer) entered in Old French at the end of the eleventh century (as *conquerre*) and derives from the popular Latin *conquaerere*. In classical Latin, there is no equivalent, but there are some paronyms: *conqueri* means "complain a lot" and *conquirere* "search everywhere." When some search everywhere, others complain a lot. This is what was found in Mexico, Peru, and all the other lands where the suitor sought to conquer everywhere.

In any event, Cabeza de Vaca had entered into a scheme that Deleuze and Guattari describe in the following way: "The nomad is there, on earth, whenever a smooth space forms which gnaws at him and which tends to grow in all directions. The nomad inhabits these places; he stays in these places, and makes them grow himself, in the sense that the nomad no less makes the desert than he is made by the desert. He is a vector of deterritorialization. He adds the desert to the desert, the steppe to the steppe, by a series of local operations whose orientation and direction constantly vary."[6] To allow these places to maintain their "open space" nature, he merely had to agree to . . . undress. And the naked conquistador crisscrossing the south of a territory (for which the Spaniards had not yet a name, but for which the indigenous population did have one, which the Spaniards didn't know about), was in the process of developing, throughout his journey, the foundations of a "treaty on nomadology." When Deleuze and Guattari wrote the premise of this treaty in *Mille Plateaux* (1980), they could have used the story of Cabeza de Vaca to illustrate the concepts of "smooth space" and "deterritorialization." Although, in Western eyes, the Texas meadows had not yet entered into a process of deterritorialization: they had not yet been transformed into "territories" by the conquistadores. But even if he was not the Cain of the Americas, Cabeza de Vaca was damned. Again, few are those who measure the extraordinary scope of his undertaking. What a pity it is for his story to be confined to the peripheral discourse of a few "Indian" history specialists, among whom I do not even have the good fortune to appear! It is 1492 that is the year of the "discovery" of a new world. Two more years deserve to be entered into the annals: 1536, which saw Cabeza de Vaca reappear among those who were his own, and 1557, the probable year of his death, which he faced with oblivion and in poverty. These are all markers of the definitive failure of a *spatial* reading of the world. Columbus had the decisive reflex, the same as all those who, after him, threw themselves into the Indies or the Americas: the world was a space that they rushed to transform into place.

The Invention of Place

In 1958, Edmundo O'Gorman unknowingly launched a tenacious trend by titling his essay on Christopher Columbus *La invención de América*. Columbus might have discovered "America" but he didn't "invent" it. Up until the end, the admiral was in the Indies. From whence came this certainty? The study of maps, of course. On a hunch, no doubt. But also from the various news stories that had not yet spread in the newspapers, but were peddled in taverns in port cities the world over: "The discovery on the beach in Galway of two corpses with Mongoloid features brought by the sea, presumably Eskimos, would have suggested to him the idea of the proximity to China, which later became one of his obsessions,"[7] confirms Thomas Gomez. To argue that Columbus discovered America amounts to feeding an anachronism, to redirecting the story based on knowledge to which the Genoese was not privy. The hypothesis of the Mexican history is already developed in my work *La Géocritique. Réel, fiction, espace*.[8]

Around 1992, several important works were devoted to the maiden voyage of Columbus and its consequences, but, as O'Gorman, the authors prefer to speak of "invention" rather than "discovery." Thus in France, Thomas Gomez published *L'Invention de l'Amérique. Mythes et réalités de la conquête* in 1992, while in the United States, José Rabasa opted for *Inventing America: Spanish Historiography and the Formation of Eurocentrism*, which appeared in 1993 (it was then translated in French under the title *L'Invention de l'Amérique*, the same as O'Gorman's essay, which for its part was translated in Quebec, in 2007).[9] In his book, Rabasa inventoried some of the terms used around the fifth centenary to describe the arrival of the Spaniards in America. *Invasion* and *encounter* were the most common alternatives to *discovery*. Rabasa himself remained faithful to the "invention" of Edmundo O'Gorman, while expressing awareness that the choice of the term could give the impression of denying "any ontological primacy of the New World."[10] To be honest, none of these terms was or is satisfactory. The arrival of the Spaniards remains *unspeakable*. Moreover, the first mention of an "invention" of America is old: it dates back to the earliest ages of conquest. In 1528, the same year that Alvar Núñez Cabeza de Vaca and the Spaniards reached the Florida coasts, Fernan Perez de Oliva brought out *Historia de la invención de las Yndias*. The Cordovan humanist, a tragedian in his day, did not explain what he meant by "invention," but in book II of his *Historia*, he delivered a striking synthesis of the project, which, according to him, Columbus had undertaken before executing his second voyage: "So Columbus, with many other men of authority who followed him, driven by curiosity to see the great novelties that he had spoken about in Spain, returned the year following the first navigation, to stir up the world [*a mezclar el mundo*], and to give to these strange lands the same form as our own [*a dar a aquellas tierras estrañas forma de la nuestra*]."[11]

In Pérez de Oliva's terms, the "invention" of place takes on a sense that refers more to the rhetoric than the technical or geographical discovery. It is about giving a familiar shape, *ours*, to something that you do not have and that is (still) *theirs*. In *L'Aventure sémiologique* (1985), an essay appearing after his death,

Roland Barthes associated ancient rhetoric with the art of making checklists and recalled that, according to legend, rhetoric was born in Sicily in 485 BC during a property trial. Aristotle attributed its first formal use to Empedocles, the Agrigentum philosopher who, it is said, one day chose to throw himself into the crater of Etna. In short, it took designing a metalinguistic device capable of supporting an argument to win possession of the land. It was only later that Gorgias, a student of Empedocles, applied rhetoric in prose speeches, including notably the threnody, the funeral speech. The terrestrial origin (*territorialisante*) of rhetoric may explain why we later persisted to "invent" lands and places. Because as Barthes points out, what is *inventio*, if not the first of five major parts of the rhetoric process, and the expression of certainty that everything existed beforehand and that it suffices simply to find it? In the vocabulary of Barthes, invention is therefore "an 'extractive' rather than 'creative' notion."[12] To invent is to gather with the greatest effectiveness the means of persuasion to support the cause being pleaded—a cause that proves to be central and whose argumentation will be used to justify or to legitimize, because these two actions are not always synonymous. *Rem tene, verba sequentur*, or "Master the subject, the words will follow," after an adage that is attributed without guarantee to Cato the Elder, to Cicero, or to Gaius Julius Victor. If *inventio* is the initial part of the process, the *actio* is the concluding part. Translated into plain language, this means that the discursive invention serves the enactment, the "action" of the recovery of what had existed prior.[13] Or, how to *invenire* on the discursive plan the manner of passing to the *actio*—that is to say, allegedly giving a familiar shape to something that you do not (yet) own legally! Ultimately, it will be to invent (*invenire*) a word destined to legitimize the coming-in (*in-venire*) of space that will compel its transformation into place.

The word and the text express the view that man holds the universe. One and the other *create* the world, locating an epiphany of signs in an intelligible framework. However, at the time of the great discoveries, the hierarchies of (hitherto) dominant discourse had been challenged. The world lost a part of reality at the precise moment when the known references, sedimented in a secular tradition, were dissolved against the general inability to explain the current amplification. José Rabasa says that "beyond the motivation of authors, novelty and reality fall within the margins of a textualized world."[14] We must now invent a new discourse that is able to account for the discovery of new measures of the human environment.

The Anthropological Machine

The extension of the world involves crossing the boundaries of human limits, the entry into a dimension in which the hero's stay is brief—that is, the dimension of the supernatural. This is generally seen as a virtual state of bewilderment. It is a threat, rather than an opportunity, to access the alternative. Faced with a geographic augmentation that results in the expansion of the human scene, the Greek myth had contributed to the domestication of this *unheimlich*, an epithet that, beyond its psychoanalytic interpretation, refers to that which goes "beyond the home." To bring into the *domos* that which was *unheimlich*, it was necessary for the Greeks to gain ground on the monstrous. We have seen how Jason and his Argonauts came to the end of the Erring Rocks and how they entered the pure space of the Euxine Bridge. This feat merits a great story. Meanwhile, deported to the remote western reaches of the universe, closer to the West, Ulysses bumped up against the ultimate limits of the territories that the gods had assigned to man. In short, he had reached the area where the human condition slipped toward the nonhuman. At this geographic end of the world, Ulysses and his men were at risk of falling into animality. More precisely, they risked losing their humanity. So they were faced with Sirens, just like the Argonauts before them. In favor of their slow return, they had in fact skirted the Tyrrhenian coast of Italy. Before weaving in and out between Charybdis and Scylla, they went around the "island covered with flowers" (Antemoessa), from which comes the fatal singing of the Sirens. In Apollonius, Orpheus managed to drown out the voice of these half-women, half-bird creatures with his lyre and a lively, fast song that charmed his companions. Despite this, Butes jumped into the water. He would have perished without mercy if Aphrodite hadn't snatched him from the waves and transported him far away from the Sirens' seduction to Cape Lilybaeum, the site of the current Marsala.[15] Butes certainly succumbed to the lure of music, but mostly he yielded to the impulse of leaving the human condition, which the uncertain geography of the places allowed. Commenting on Butes's gesture, Pascal Quignard writes, "He left his rank. He climbed the prison wall. He joined the sovereign spontaneity of nature." And he emphasizes "Butes' rush towards an anterior animality."[16]

Facing the Sirens' island, none of Ulysses's companions would commit such imprudence. Could it be that wax in the ears was more effective than Orpheus's song? It is therefore that from one generation to the other the world was technologized. For Ulysses, the danger comes from elsewhere. In song X of *The Odyssey*, Circe on her island poetically turns men into swine; she marks the end of the transition process between the human and that which is no longer entirely human or is no longer human at all. Thanks to the moly, the magical herb that Hermes made him pick, Ulysses resists the spell. He masters the goddess and incites her to free his companions: "They became men once again, but younger / and much more beautiful and taller than they were before."[17] When the transition is complete, it is as if the newcomer rediscovered his humanity, but he is reinforced by the achievements of an experience that allowed him to increase his potential. He's younger, more handsome, and taller. He stretched the limits of his world, and he overcame the boundaries. Ulysses and his crew foiled the trap that could have allowed Circe to transform them into monsters or, more simply, into beings with the head, the voice, and the bristles of pigs. They managed to tame the provisional rim of the world, to push the monstrous a little further away. The familiar universe has expanded to the extent of this conquest, although in Homer it is fragile. Without the intervention of the god Hermes, it is a safe bet that Ulysses would have succumbed to the spell of Circe. And there is little doubt that the light-footed Hermes would not always give his support to man, to the potential conqueror—a status Ulysses assumes for the role he played in the capture of Troy rather than for his maritime exploits, which were almost incidental.

In *L'Ouvert. De l'homme et de l'animal* (2002), Giorgio Agamben spoke of an "anthropological machine" that, working in the culture of a society (Western, here), aims to produce the human through a set of oppositions contrasting the human and the animal or the human and the inhuman. This machine carries a tautology, since the human, a quasi-divine figure, is always predetermined. Therefore, it creates "a sort of state of exception, a zone of indeterminacy where the outside is only the exclusion from an inside and the inside, in turn, is only the exclusion of an outside."[18] A version of this machine belongs to the moderns and the other to those of antiquity. The one of the moderns tends to exclude some humans from the human sphere by animalizing them. That of antiquity functions in a diametrically opposed manner:

> If, with the machine of the moderns, the outside is produced by the exclusion of an inside and the inhuman by animalizing the human, here the inside is obtained by the inclusion of an outside, the non-humans by the humanization of an animal . . . Both machines can only operate by establishing in their center a zone of indifference where there must occur—like a missing link which is still missing because it is already virtually present—the interaction between human and animal, humans and non-humans, the speaking and the living. Like any exceptional space, this area is, in fact, perfectly empty, and the truly human which should come there is just the place of a constantly postponed decision where the ceasuras and the rearticulations are always again dislocated and displaced.[19]

To this author's mind, the reflection concerns the paleontological discourse of the late nineteenth century (Ernst Haeckel, Heymann Steinthal) and the question of the origin of language, which one wonders if it, like laughter, was man's own or if it was the result of a type of Darwinian evolution. But the anthropological machine is likely to find many applications in the most varied fields. One of them rather closely concerns the mechanism for fixing place. It goes without saying that Agamben's concept presents a strong analogy to the "war machine,"[20] the enemy of nomadism, that Deleuze and Guattari described in *Mille Plateaux*. The anthropological machine is set in motion whenever the individual is confronted with a new space, which is confusing and which he must just as soon enclose. It helps him understand the area of ontological indeterminacy that arises before him. It helps him transform that which is dislocated (*dis-locatus*) into a controllable place (*locus*). It performs an integrative function. Brutally integrative. It immediately precedes the invention—that is, the transformation of an indefinable space into a common place.

At Aeaea, Circe's island, did the familiar world of man really expand? Maybe not in the shaky cosmographic vision of Homer. Did man have his place there where the sun sank into the sea? After all, only a god had allowed Ulysses to visit the end of the world without getting lost. His adaptation to the teratological nature of the absolute West was short-lived. Certainly, he now knew the antidote that would protect him from dehumanization. But nothing could stop one from thinking that Circe could change her poison, if she had so wished, and that Hermes could refrain from further intervention. Whatever the degree of exceptionality, the adventure of Ulysses and his crew teaches us that man is able to resist the monstrous nature of open space. But Ulysses accomplished a rare feat, which is perhaps unique. He did not content himself to just triumph over the outside monster (which he had already done by overcoming Polyphemus on the Island of the Sun); he surpassed his own monstrosity, the internal monstrosity that going to the limits clearly evidenced. *The Odyssey* gives us yet one more lesson. Song X tells how man was able to survive crossing the natural boundaries of the world; so that which was once unspeakable was now captured in words. What has been done once may not be repeatable; what has been said can always be repeated. The invention of the world unfolds in whatever world, whatsoever: a possible world, to which the real world will eventually adapt.

Battling monsters was the true indicator of the interplay of limits throughout the medieval period. The north and south served as theaters to the most spectacular performances, according to a logic based on the theory of climates. In the imagination of the Middle Ages, *north* and *south* referred to extremes regions. Sometimes it was terribly cold, sometimes too hot, even though such extremes are to be avoided. Around 1410, Pierre d'Ailly explained this in his *Imago Mundi*, a cosmographic treaty that became an authority throughout the century and influenced many. Columbus scrupulously annotated the margins of the copy he had. According to d'Ailly, "there are four extreme regions located within areas that are habitable, and of the four it is mainly two of these regions, one of which faces the south and the other north. This is why Ptolemy, Haly and others claim that in these two extreme regions there are cannibalistic wild men, with deformed and horrible faces. Haly attributed this phenomena to the unequal distribution of heat and cold in these regions, the cause of abnormal complexions and hideous deformations, and also the cause of perversion and coarse language: these are beings which are difficult to distinguish between men and beasts, in the words of blessed Augustine."[21] In this way the mysterious biblical couple Gog and Magog (who became Yajouj and Majouj in the Qur'an [18.94, 21.96]), expresses an evil principle that seems to contaminate the good northeastern edge of many maps, such as on the Hereford mappemonde or that of Juan de la Cosa. Mercator did not forget them but rather relocated them to the northeast of Siberia where the *ultima tellus* has been moved to at the end of the sixteenth century. To the south, there are men-birds, or beings without ears, who tend to indicate the last worlds. According to the principle so dear to Homer, when humanity broke down before animality, it is an indicator that the ecumene was losing its structuring force in favor of the hybrid. Africa in particular bears the brunt of this demeaning dynamic. Mercator calmly explained that because of the dryness, animals congregate at watering holes where they copulate before giving birth to monsters. Climate theory is influenced by genetics! Even earlier, Pierre d'Ailly, who was the cardinal of his state, had inhabited the continent of baboons—that is to say, men with heads like dogs, who expressed themselves by barking like dogs. For some, it is the baboons of the religious cosmographer that inspired the cannibal character of Columbus, who was a faithful reader of Ailly. The navigator understood Latin better than the language of the Arawaks. Did he misunderstand his interpreters, who for their part, were supposed to have learned Spanish, the language of the aliens, in just a few weeks, or even a few days? *Can* (or dog, the man with a dog's head), *Khan* (the ruler of the place that Columbus believed to be approaching), *caniba* (or the abstract name of a tribe, the Caribbean) . . . There was room for confusion, indeed. But the surest way to establish communication is always to tackle the known (or the fantasized) about what you do not understand. One can imagine the consequences of the process. In the landscape of the New World, there is a man with the head of a dog that eats the flesh of his fellow beings. The West had temporarily taken the place of the South in the discourse on boundaries, which is always a discourse that renders itself delirious[22] in order to escape its own limitations.

Monsters kept guard for a long time there where man risked losing his humanity. The Renaissance did not break with this antediluvian tradition that was conveyed by the majority of cosmogonies; it even strengthened it. The tradition continued to propagate headlands and other straits in teratological geography. As it became more difficult to see Charybdis and Scylla in the Strait of Messina or the Erring Rocks blocking the Bosphorus, the monsters were displaced at a respectable distance, to new extremes. The Cape of Good Hope was circumvented by Vasco de Gama in December 1497, after his compatriot

Bartolomeu Dias, the "Captain of the Confines,"[23] discovered it in 1488. For the latter, it had become the Cape of Storms. These waters had instilled fear in him. But King John II was optimistic and sure of himself, to the point that he had rejected Columbus's offers of service. He had realized that his sailors had just opened the route to India by the south. The name changed as well: "Good Hope" filled in for the "Storms." In 1652, the site became a Dutch colony. As for the Khoikhoi (later dubbed the Hottentots) who inhabited these lands for the last thirty thousand years, little was said. Halfway between these two dates, Camões sang in his *Lusiades* (1572) of the exploits of Vasco da Gama, narrating the passage through the cape. He even referred to the Khoikhoi, but the role he assigned them was not very original: that of the savage. Located to the south of a region that remained mysterious at the time when Camões wrote, the Cape of Good Hope could not content itself to be just a simple stone promontory. It therefore became a monster. In song V, the Portuguese followed along the coast of Africa and, reaching the southern tip of the continent, entered a huge black cloud. Dread seized them while they were in a threatening and open space, when a gigantic figure appeared vertically, with a deformed face, unkempt beard, sunken eyes, hair covered with mud, a dark mouth, and yellow teeth. It is Adamastor, a survivor of the gigantomachy in which Zeus and the Olympians triumphed over the forces of Chaos. Coupled to the rock like Prometheus once was to the Caucasus, the titan Adamastor is the eternal guardian of places that nobody has ever glimpsed. But he is already resigned. He is resolved to see Vasco da Gama overcome prohibited barriers in order to dominate the vast eastern seas. Not the slightest storm is triggered, not the smallest pebble is thrown into the water. But, unlike the Sirens of some later versions of the *Argonautiques*, the consciousness of failure does not lead to suicide. The rock does not collapse. Adamastor holds back from delving into the southern emptiness. Nothing happens, nothing at all. Obviously, the monsters are no longer able to impede the progress of an expanding Europe. But they still have the gift of speech and prophecy. Flood control is within reach of Western man, but the control of time and the future is not. Adamastor, who knows the future, foretells of the worst reprisals. This time, he will keep his word. The Portuguese die not very far away. Because, as he himself says, "I'll end here all of the African coast, with my promontory never being seen, which extends to the South Pole, and which is there to offend your temerity."[24]

In 1927, a statue of a titan was erected on the Santa Catarina lookout, in Lisbon. Jean-Yves Loude had reached that point at the beginning of the new millennium. Adamastor has finally been tamed. He has even been emasculated: "If Adamastor was complaining so much, it is because his genitals had been cut off."[25] The Titan is the rock on which the conqueror built his colonial temple. But this animated stone feels captive, humiliated. In Loude's beautiful prose, "Adamastor heaving, breathing, roaring, a failed King, his secrets bare, the extremity of the African body brought back in chains. Pass over his curly head, rough stone, less that polished, a helicopter, a plane; the train growls at his feet; boats whistle mockeries, they know that the way is clear, that the cape is defeated, that the monster has been retained for a long time in the nets of order, around the north."[26]

An amazing poetic creation of Camões, Adamastor did not spoil Dante's *Inferno*. In the production of monsters, the Age of Discoveries was moreover engaged in a process of selection that allowed the anthropological machine to renew itself. The distance is shortened between man and monster: the de-monstering/demonstration of Adamastor proves it. By dint of reducing this distance, we ended up giving a new face to the monster: a face with feminine features! The woman was initially one of those Amazons, of which the sixteenth century encouraged the return. The Amazons have always been accustomed to populating the margins of the world. In the northeast, they had cleared a path through the steppes of Scythia (Herodotus) until Gog and Magog followed their footsteps and eventually took over. To the southeast, they were established in their city Themiscyre, at the edge of Thermodon, in Asia Minor (Strabo): that is where Hippolytus was stripped of his belt by Heracles. Confused with the Gorgons, they were also transformed into warriors in Libya. In the *Periplus of Hanno*, they inhabit a nested space in the deep south: an island in the heart of a lake, which is itself located on an island. They have a hairy body. According to Giovanni Battista Ramusio, who does not forget Hannon in his compilation of the great navigations of the past, an interpreter had explained to the Carthaginian sailors that they were gorgonians (*gorgónes*). But Ramusio is wrong: the Greek text has the interpreter say that they are *gorillai* and not *gorgónes*. The term had no meaning for the Italian scholar, but he was destined to experience a great success from the mid-nineteenth century after the American naturalist Thomas S. Savage named the big ape *gorilla*, having identified its bones during a stay in Liberia. There is evidence that Savage had read an English translation of *Périple* in the eighteenth century. Perhaps he did this reading during his voyage to Africa. What fate did Hannon reserve for the *gorillai*? He had three skinned in order to bring back the pelts to Carthage.[27] He was the new Perseus who had come to the end of a Medusa revisited. The monster had already become vulnerable, but he continued to indicate the end of the world. Upon arriving at the *gorillai*, the Carthaginians turned back for fear of running out of supplies.

The three known regions in the ancient world are therefore barred by women who are more or less dangerous, more or less feminine.[28] America would not be an exception. Along the way, we would come across the Sirens, half-woman half-bird, half-woman half-fish. But the land and the seas were now silent. There are Amazons everywhere. They can still be seen today. What do you think Hippolyte did after being defeated by Heracles? She took her misfortune patiently and one day moved to New York where she engendered *Wonder Woman*, William M. Marsto's famous comic strip heroine, popularized in the seventies by a bad television series. Before the era of media, when Amazons could not be seen, they were guessed at. In 1595, Sir Walter Raleigh undertook the mission to discover Eldorado in the region of the Orinoco River. He failed, as Aguirre before him. Upon his return to England, he published *The Discoverie of the Large Rich, and Bewtiful Empyre of Guiana* (1596), in which he told about his investigation of the Amazons among the locals. Did he also resort to using interpreters? In fact, the Amazons were richer than they were disquieting. They had the ever-coveted gold, which they procured in exchange for spleen stones, green stones, and jade. But the Amazons always frolicked a little further away, like their gold—like Columbus's gold, which was still shining on the next island.

It was not, however, Sir Walter Raleigh who made of this vast expanse the "Amazon." The honor had fallen to the Spaniards. The English poet-pirate was preceded in 1541 by the conquistador Francisco de Orellana. Leaving from Peru, Orellana came across the Icamiabas on his way, who, according to local tradition, were warriors. Impressed and almost defeated by these formidable opponents, he gave the river he descended the name *Amazon*. Much further north, it was still one of these warriors who permitted the Spaniards to fix both the limit of the Indies and a name that was destined to resist time. Princess Calafia reigned, indeed, over California, an insular kingdom of black Amazons whose weapons were made of gold and who lived in contact with griffins and other strange creatures. This kingdom was located, it was thought, in the current Baja. This peninsula was located in 1534 by Fortún Ximénez's men, who during a mutiny killed Diego de Becerra, a conquistador sent by Cortés, but continued their journey northward anyway. In fact, this kingdom appeared out of Garci Rodríguez de Montalvo's imagination around 1500, who himself was author of *Las sergas Esplandián* ("*The Exploits of Esplandian*") and who continued the famous *Amadis de Gaul*. For it was in search of the Amazons that Cortés had sent Becerra—the Amazons and, therefore, gold. Cortés read, and he lent credence to what he read. California was considered to be an island at the time, and it long retained this feature (sometimes until the eighteenth century). The Amazons appeared everywhere, all over the world, each time milestones were planted. Their popularity could be measured by the chivalrous romances that they animated and that were devoured during endless Atlantic crossings or long tropical evenings. Sometimes they even worried residents of the West. In his essay, Thomas Gomez has unearthed a letter that Martín de Salinas, an official in Valladolid, sent in 1533 to a secretary of Charles Quint. He reported on a rumor that was going around the Iberian Peninsula: people had witnessed the landing of ten thousand Amazons in Santander and in Laredo, where they had planned to get knocked up for the handsome sum of 15 ducats, thus provoking a crisis in the Cantabria ports' prostitution ring.[29]

The anthropological machine continued its work in the sixteenth century. Although there were a few properly monstrous women, it was more about monsters presented in feminine forms. The Amazon became more feminine and started to look like an Indian. Without necessarily being seductive, she became susceptible to seduction. The irresistible conquistador would have used his power of fascination over subjugated natives to advance overseas the cause of the crown! But in the dominant discourse, the woman remains a woman, especially when she is an Amazon converted into a native of the West Indies and the man is misogynistic and utterly convinced of his macho superiority. In the Strasbourg edition of Vespucci's *Voyages*, dated 1509, an engraving represents a robust sailor who was sent to earth to seduce women and coax them. All would have gone according to plan if one of them had not treacherously placed herself behind him in order to knock him out. A second engraving renders a sinister verdict. While the foreground describes a peaceful village scene, in the background a woman holds a hatchet above severed limbs—those of the sailor, without doubt.[30] It is far from the beautiful love story that, a bit later, brought together Pocahontas and John Smith! It is more on the lines of the rather dangerous liaison Ulysses had with Circe. When Hermes indicated the moly to his protégé, he gave him interesting advice:

> I will tell you of all the wicked witchcraft that Circe will try to practise upon you . . . When Circe strikes you with her wand, draw your sword and spring upon her as though you were going to kill her. She will then be frightened and will desire you to go to bed with her; on this you must not point blank refuse her, for you want her to set your companions free, and to take good care also of yourself, but you make her swear solemnly by all the blessed that she will plot no further mischief against you, or else when she has got you naked she will unman you and make you fit for nothing.[31]

Homer was the first to describe the mixture of violence and sexual ambiguity that created the zone of preliminary turbulence in which man faced the feminized space that he strove to turn into place, into the marital home. He "strove," yes, but in striving to do so he exposed himself to losing his virtue.

The episode that cost Vespucci's sailor his life is unfortunate, but it hardly slows the process of eroticization in which those who replaced the guardians of traditional boundaries were embroiled, the models of which were offered by Greek mythology. If Vasco de Gama had exceeded one monstrous interdiction in order to enter the "East Sea," Magellan did not ask anyone anything in order to force his way through the strait that would later bear his name. Somehow, the southern tip of America is like the practice field of the male conquistador, where all kinds of phallic symbols intersect. This sexual reterritorialization has been carefully described by Ricardo Padrón in *The Spacious World*. For him, "Magellan arrives to make an honest woman out of the land he penetrates and names."[32] In literature, this geography of sexual desire has undoubtedly found a privileged expression in *La Araucana*, published three times between 1569 and 1589. Alonso de Ercilla y Zúñiga, who had participated in 1557 in a bloody war against the Mapuche (the *araucanos*, or Araucanians), delivered a truly epic literary version of the conquest of Chile, which he passed off as a native uprising. According to Padrón, Ercilla had understood the "penetrative consummation of a relationship between the explorer and the land."[33] It is true that due to its elongated shape, phallic so to speak, the map of Chile is ideal for symbolizing male desire. I'm not sure that the conquistador was aware of this feature: Freud had not yet been born, and maps were rudimentary. But the quasi-erotic character of the conquest does not seem to have escaped him. In Padrón's opinion, he even devised a "subtle counter-cartography of imperial desire."[34] Experiencing a rare sympathy toward the Araucanians and the young cacique Lautaro, Ercilla did not stop valorizing the bravery of an enemy whose resistance seemed legitimate. Also at the heart of the Indies, Ercilla recognized the greatness of the Other, the "barbarian" (as opposed to the "Christian"), even before Montaigne composed his famous panegyric of the Noble Savage in chapter 31 of book 1 of his *Essais*. When Ercilla depicted the battles in which the Spaniards and Araucanians tore themselves apart, he insisted on the beauty of the nearly naked opponents' bodies. The fray, albeit of the worst possible violence (the warriors gleefully mutilated each other), was transformed into a quasi-romantic embrace. In song XV, Andrea, an Italian soldier, is engaged in a fierce dual with the Araucan Rengo. The blows rained; they were dodged; they were returned; the opponents were entwined: "The mighty barbarian seized him, and, their bodies pressed together, he squeezed him with his bare arms. The chainmail was imprinted on his chest. The Lombard does not fear; he hopes to finish off the hero more easily: he hugged him with his arms covered with steel and believes he can raise him off the ground."[35] Rengo pushes. Both eventually die. It's a fight of arms coated with steel versus bare arms. The Araucans will be submissive, like the Amazons. The Araucanians had women who, according to Ercilla, had little dignity: they really let the dying soldiers have it, making them "die a thousand ways; because a cruel woman is cruel without measure."[36] This is the message that the illustrator of the Strasbourg edition of Vespucci's travels tried to communicate.

Space was locked up by monsters. A significant number of monsters were feminized. To monopolize the space and make it a place, to *dar a aquellas tierras estrañas forma de la nuestra*, it was necessary that man should come to the end of this woman who was dangerous by vocation and sometimes seductive, more desirable as her dangerousness declined. "As the naked body of the Indian, the body of the world becomes a surface offered up to the inquisitions of curiosity,"[37] Michel de Certeau noted. The allegory would explain the process that conquering the extreme had begun. Africa and Asia were presented as feminine in the time of Herodotus, and Europe was too. Asia was an Océanide married to Prometheus; Africa was initially Libya, granddaughter of Io and mother of Agenor; the latter was the father of Europe. The myth of Europe is well known. Etymology places Europe in the genealogy of Libya and, as she was a princess in Sidon, also in the geography of the East. Libya is the Africa of the Ancient Greeks. The etymon of Africa is more recent; it probably comes from the wind *Africus*.[38] But Africa remains feminine. The three classic continents are the objects of feminized representations. It goes without saying that America would not have known how to be an exception to the rule: it would completely become a naked Indian served up to the avid gaze of the explorer (and therefore the gaze of the conqueror). In this context, *America*, an engraving by the Flemish Jan van der Straet, better known under the name Stradanus, drew the attention of Michel de Certeau and José Rabasa.[39] Toward the end of the sixteenth century, *America* recounts the arrival of Amerigo Vespucci.

A native is lying in a hammock suspended between two hardwood trees; she is surrounded by exotic animals; in the background, as in the illustration of 1508, two women are busy roasting a leg over a fire. The beautiful indolent seems to be speaking to a man dressed in the garb of an ideal navigator, sextant in hand, a caravel boat within reach of his rowboat. Americus has just met the lady who will take the name of America, as a wife inherits her husband's surname. There is little doubt that the wedding does not celebrate a marriage of love. Rather, it sanctions the seizing of possessions. "What begins as such," says de Certeau, "is a colonization of the body by the discourse of power."[40] And what is confirmed is that the anthropological machine has passed from its old version to its modern version, if one follows Agamben's grammar. Now it is no longer the companions of Ulysses who are transformed into swine. The Western man, even to himself and as an absolute reference, no longer runs the risk of his humanity dissolving into an uncontrollable animality. Rather it is the Other who is transformed into an animal—but into an animal that is still human enough so that there can be the question of taming it, of "domesticating" it, of transforming the confusing space it inhabits into a domestic place. In a Deleuzian language, the smooth space of the Other is transformed into the striated space of the Same. In a Western society dominated by males, this passage is translated as the feminization of the iconic monster; the woman is perceived as the ontological Other *par excellence*.

For a while, things stayed that way. In 1662, in the planisphere that opens the *Atlas major* of Joan Blaeu (a worthy scion of a family of Dutch cartographers), a naked, feathered woman is lying in a lascivious position, her head resting on the palm of her hand, as if she is waiting for a new explorer. The chariot that supports her is placed on the edge of the South American continent. And California continues to remain an island. Then the situation changed. In the 1840s, some Americans were convinced that the "manifest destiny" (or the divinity), of their nation led them to conquer the West in order to spread civilization—that is, democracy. They began to make war on Mexico. And what was not taken by force was bought or negotiated (Oregon negotiated for with the British, for example). But it was not just the West. The South was also targeted. The Central American adventure of the filibuster William Walker was part of this ideological framework. A new feminine allegory was used. The woman was not Indian or lascivious; she had no hammock or pen; she was not expecting anyone. Now she floated in the air, virginal, dressed as a vestal, and guided pioneers' wagons through the vast prairies; she drove away the darkness and brought light while the bad "Indians" were fleeing toward the obscured part of the horizon.

In her right hand she held a textbook, in her left hand, the telegraph wire. She was blonde. She was a Yankee. This image is Columbia, a personification of the United States of America. This is how she appeared in 1872 in *American Progress*, a patriotic painting by John Gast. Today, one would expect Columbia to crisscross the skies of Iraq and Afghanistan. To distill democracy remains the preferred form of fulfillment of a destiny that the medias render more and more evident, but now F-16 fighters occupy the airspace of feminine allegories. This is probably because the graceful Columbia was forced to give way to the first great male allegory embodying a place: Uncle Sam, who was born, already graying, on a poster conceived by James Montgomery Flagg, on April 6, 1917. *I Want YOU for U.S. Army*. Everybody is familiar with his threatening finger.

Was the new monster decked out with a masculine nature? A clue comes once again from the confines of the world from which the Amazons have disappeared. In the early sixties of the twentieth century, Peregrino Fernandez agreed to lead the small team from Confluencia to Cipolletti, Patagonia. Peregrino Fernandez seems lifelike, but he is a character created from scratch by Osvaldo Soriano. Before becoming one of the great writers of his country, Soriano was a confirmed *goleador* in southern Argentina. Several of his short stories about soccer testify to this. But maybe they evoke his sports *Doppelgänger*. For him, storytelling is king. In his first novel, *Triste, solitario y final* (1974), he represented himself helping Philip Marlowe with an investigation into the last months of Stan Laurel's life in Los Angeles. As for Peregrino Fernandez, *El Míster*, the enigmatic coach, he punctuated workouts with quotes from Schopenhauer and knew, among others, that "defenders always seem bigger and uglier than they are because they indicate to us the limits."[41] You have to read Schopenhauer to understand that the closer one gets to the boundary, the more monstrous and scary the human seems. The monster had readopted a masculine aspect, that of the stopper, one of the men from the center back, the one that bars access to the penalty area, 16 tiny yards from the goal, from the Eden of modern times. And it is the center forward's responsibility to know his way around this new Adamastor, now rampant in all fields of the world, in Argentina, in Europe, even in South Africa. Ah, the memories of Van Bommel's (the Oranges's worrying midfielder) grim look, fixed on the forward of the 2010 World Cup. No need, however, of bulky caravels in order to overcome the monstrous obstacle; just as effective is the swaying dribbling of the winger. The forward is a small-scale adventurer. Peregrino Fernandez advocated for increasing the number of forwards. Upon his arrival in Cipolletti, only a fullback found favor with him; the others had to yield to the wingers and the midfielders. Matches often ended with scores of 8 to 6, for the opponent. Bypassing the goalie, big and ugly, the rusher opens space, instilling in the spectator a sense of liberation. Footballers are more popular than the Argonauts in their time, yet they share some of their fears.

The Two Senses of Invention

In rhetoric, invention participates in the development of an argument that ultimately aspires to produce something hitherto unseen. Invention is based on actual fact but claims to innovate, introducing a variable that, already in the original judicial context, should take aback the adversary. The invention of place stems from the observation of the existing, but does not always aim at the emergence of a new meaning. Sometimes, the goal is to fit the new into the old. The invention of place in the fifteenth and sixteenth centuries has almost invariably led to the reclassification of fact or the state of things past. We refer to an *exemplum* that serves as a type of model, a referent, to invent by inference—at best, an indicator or a sign is used to engage in a deduction leading back to the known. Rhetorical topos and spatial topos merge to build a common place, both banal as obviousness and reassuring as everything that is shared.

In *La Géocritique*, I did, like many others, make the comparison between Jason's or Ulysses's trips and the first Greek expeditions directed toward the Euxine and the western Mediterranean. To be honest, Jason and Ulysses were themselves not the first. Planned for fifty rowers, the ship Argo was to make of Jason the first man to brave the high seas. But what about the "many ships" that Aiétès, King of Colchis, father of Medea, had launched on the heels of the Greeks' penteconter? The answer is obvious: the Argonautic primacy was relative. When Ulysses and his men skirted the island of the Sirens, were they the first? They had been preceded by the Argonauts, but the fatal islanders had forgotten: "Never had a black ship sailed around our cape."[42] And the Greeks had plenty of time to check and see if the warning that Circe had sent them was founded: the island was littered with bones and corpses whose skin was drying out in the sun. The travels conveyed by literature, which for a long time were not classed in the fiction section but in the broader section of narratives informing the world, served as laboratories of possible material explorations. Has the situation changed much? Certainly, the primacy of travel was argued, but literature continued to take inventory of the more or less prestigious predecessors. Readers still wanted to be reassured; the mythical and legendary figures were still mobilized to corroborate a founding presence, a territorial precedence claimed against rivals on the lookout.

Sometimes the stories shamelessly exploit myths. Gonzalo Fernández de Oviedo provides a good example of wishful thinking in his *Historia General y Natural de las Indias*. According to him, the "new" land had not been discovered but simply rediscovered. He finds that "these lands had been forgotten."[43] Once, they figured on the list of the possessions of the king of Spain. Citing an Aristotle revisited by Theophilus de Ferrariis, an obscure priest of Cremona that posterity has spurned, Fernández de Oviedo said that Carthaginian merchants had, once upon a time, implanted themselves on an island characterized by a "high fertility" (*grande ubertad*).[44] Faced with the threat of depopulation of the motherland and the fear that others would seize these dangerously prosperous places, the Carthaginian senate forbade anyone, under threat of the death penalty, to take the path to the island. It was soon forgotten. The stage is set: scholarship is positioned in the service of the invention of place, itself guided by ideology. But the rich imagination of the historian did not necessarily alight on the right path. Invoking the authority of Berosus the Chaldean, a Babylonian scholar, and that of the Archbishop Rodrigo de Rada, who had composed a chronicle on the glorious past of Spain shortly before the middle of the thirteenth century, Fernández de Oviedo claimed that the Carthaginians were not the first. Before them, the island was associated with the Hesperides, well known to the Ancient Greeks and the mythologists. From where did the Hesperides get their name? From Hesperos, the Greek evening star, which became Vesper to the Romans? Wrong. From Hesper, rather, who was the twelfth (legendary) king of Spain. To all those who dare doubt Fernández de Oviedo's explanations, they can consult the *Etymologies* (IX, 2). In this text, Isadore de Séville recalls what the etymon of peoples was after the fall of the Tower of Babel: the ethnonyms derived from the names of great figureheads—kings, most of the time. According to the principle established by Isidore and the interpretation of our chronicler, Hiber, the second king of Spain and descendant of Hercules, gave his name to the Ebro and to the whole of (H)Iberia. As for Hispania and Spain, they owe theirs to Hispanus, another master of the place. Not to be outdone by the Italians, who said they were descendants of the Trojans, the Spaniards were not going to simply settle for a bunch of Trojan ancestors. They claimed the honor of being themselves the ancestors of the Trojans! Brigus, the fourth king of Spain, gave his name to the Brigiens, who were the ancestors of the Phrygians. It goes without saying that the paronymy Brigiens/Phrygians seemed obvious to Fernández de Oviedo. These Phrygians from the Brigiens would later become the Trojans, of whom the Iberians are the predecessors. We're still wondering to whom belongs the intellectual property of Marianne's Phrygian cap![45] Ultimately, the trip to India is presented as a return to the Hesperides, to which the twelfth king of Spain gave his name 171 years before the foundation of Troy, which based on the calculations of the author, intervened in 1487 BC. Advanced by five years, this dating would have resulted in quite a nice symmetry. Think about it—Troy founded in 1492 BC! The Americas of some are the *Indias Hespérides* of others. For what it is worth, I note that the position of these *Indias Hespérides* coincided, according to the chronicler, with that of Cuba and Hispaniola. And the conclusion imposes itself: "God gave this seigneury to Spain after so many centuries."[46]

Much more recently, in 1996, José Manuel Fajardo has published a novel titled, in French translation, *Lettre du bout du monde*.[47] It recounts the events that occurred in Hispaniola after Columbus had left on his return trip. Thirty-nine men remained on the island in order to perpetuate the Spanish presence, to convert some of the Caribbeans, and to build a fort. In fact, they had had little choice. As the admiral had damaged one of the caravels, there was not enough room on board the remaining two for everyone to go home. The fate of these men is known: none of them survived. What the official story does not say, however, but what the Grenadian novelist suggests to the reader is that Domingo Pérez (a cooper by trade, who was part of a small group) kept a diary intended for his brother. It is learned that the Spaniards, driven by the thirst for gold and torn by increasingly deep dissension, disperse themselves and kill each other when the opportunity arises. They would not fail to treat the First Peoples with cruelty. We would also learn that a supreme leader with white skin, to whom the natives ascribed divine virtues, reigned in the depths of the island. It was Yucemí, the "White Spirit." As it should be, Yucemí is the guardian of a treasure. Domingo Pérez meets him by chance in a cave. Their encounter is neither warm nor banal. Domingo Pérez and the few men who accompany him are indeed shouted at in Portuguese and called "slaves of gold." Columbus's cooper did not take long to understand that Yucemí was a Portuguese sailor named Alvaro Almeyda, who landed on the island 15 years earlier with two companions who had since been killed by the Caribbeans. Sick with the plague, Almeyda himself is dying.

This story is not straight out of the imagination of José Manuel Fajardo. Several sources converge prior to the novel to confirm it. But that is not all. Fernández de Oviedo was only *one* of the great chroniclers of the Indies. There have been others. In 1552, Francisco López de Gómara brought out the first edition of his *Historia general de las Indias*, in which he began by refuting the hypothesis of his compatriot. No, the Hesperides were not located in the Caribbean; they were off the coast of Africa, more precisely between the Cape Verde Islands and the Gorgons. But, in his own way, López de Gómara also argued in favor of a "rediscovery" of America. In chapter 13 of his chronicle, maliciously titled *El descubrimiento primero de las Indias*, he reported a story that caught great attention thereafter. Before 1492, an anonymous caravel, driven by easterly winds, entered far into the Ocean Sea. Upon its return to Spain, most of the crew died of hunger and exhaustion, with the exception of the captain and three or four sailors. The latter would die shortly after arrival without having said a word. As for the captain, he survived a few more days. What was his origin? The chronicler is not certain: Andalusian, perhaps, or Basque, or Portuguese. Fajardo chose to make him Portuguese. But this option was not necessarily the one that attracted favor in the sixteenth century because it had to be shown that it was certainly a Genoese who had "discovered" America, but he did so within the still frothy wake of a Spaniard. Despite the numerous versions of the legend, one aspect remains invariant in all cases: the pilot was received by Christopher Columbus in his home. On this occasion, the future pioneer of transatlantic travel had plenty of time to browse the logbook of his guest, to be seduced by his reading and to memorize the coordinates of the new land.[48] Fernando Iwasaki summarizes the situation with humor: "So Columbus had touched land in 1492 which had already been mentioned in *Genesis*, which appeared in the Plato's *Dialogues*, that the Phoenician traders had exploited in the days of Solomon, that Pliny had described in his *Natural History*, which were evangelized by the Scriptures, and which an unknown Spanish captain cruised before an upstarting Genoese arrived, brandishing a license to discover. 'What a colonist you are, my little chick,' my mother would have said to the face of the Admiral."[49]

The national or nationalist fiber vibrates across all the stories. It would take a long time to establish their complete inventory, but it would be amusing. For discoverers and their compatriots, the primary concern is to establish the precedence of their presence. Myth underpins ideology. We attribute neighborhoods to ancient nobility. How many people, including the Franks who are considered to be descendants from Francion (the apocryphal brother of Aeneas), have the unfortunate Trojans named! But the nationalist thrust is not the only driving factor in this process. The modern nation and the exclusiveness (or exclusions) that it attaches to legitimizing are recent, and yet the tendency to make every original voyage the twin of a yet another original voyage is old. The retraction of the inaugural character of exploration in favor of a first mythical, legendary, or anecdotal discovery (*el primero descubrimiento*), demonstrates once again the willingness to deprive space of that which is consubstantial: the immeasurable. If we are not careful about containing the spatiality of space, we will make traveling into the unknown a passage on a tightrope, upright acrobatics over emptiness. Faced with this appalling prospect, claiming the virginity of that which is penetrated stays in the background. The less the land is virgin, the more reassuring it is. The phallocracy surrounding conquest is a cowardly regime. Allaying fears justifies all concessions. It is at the time of Spaniards' massive installation in the Indies, and others elsewhere, that the doctrine of the *mentalis restrictio* has flourished under the leadership of Martin de Azpilcueta, a prestigious professor from the University Salamanca, known as the Navarrus. God becomes the sole witness of *all* human truth; as for men, it suffices, if you will, to reveal to them a mountainside—or how to increase the opening of the world and create an expression minus the truth. For example, can a fear of heights be concealed without breaking God's laws? Navarrus inspired the Jesuits who knew so well how to combine composition of places and mental restriction to draw the world in the image they so wished. *Ad majorem Dei gloriam*, according to the motto of the Company of Jesus.

Name of the Country: The Country

In order to take position of a *place*, the *space* must be emptied of its spatiality. That is the key word. And what surer way to achieve this purpose than to attribute a new name, a toponomical baptism. This act anticipates another baptism, that of the "native" to be converted, which in the meantime will be shown on the landing beaches, the d-day case *without* the capital letters of the discoverer. If the space is perceived of in an original and illusory virginity, the place is usually inhabited. This area is often the location of Another who is not yet identified. Therefore, the abstraction of the idea of space encounters the concreteness of place, which is embodied in its first occupants. The shock was accompanied in most cases by great brutality. Millions of "natives" did not survive. While Europe inaugurated modernity in initiating the colonization of a "new" world, its people suffered the brunt of what is called in the Quechua language the *Pachakuti*. Walter Mignolo spoke about it in *The Idea of Latin America* (2005). As he explains, *Pacha* corresponded to *Gea*, the "Earth" (others translated the term by the conjunction of space and time), whereas *Kuti* indicates the dramatic change in the order of things. Therefore, for the Taino people of the Caribbean, as for the inhabitants of Tawantinsuyu in the Andes and for those of Cemenahuac in the valley of today's Mexico (examples given by Mignolo), the arrival of Columbus, Pizarro, and Cortés was a *Pachakuti*: "a violent destruction, relentless invasion, and disregard for their way of life—a convulsion of all levels of existence."⁵⁰

The title of my paragraph pays tribute to Proust, whose protagonist, failing to really travel about, resigned himself to seeing the country only through place names, in the names of countries. It is not so much the author of *La Recherche* that I am thinking about here, but rather Stefan Zweig. Not so much Balbec as Brazil. Zweig did not survive the intrusion any more than the Tainos. But the one he underwent occurred five centuries after the arrival of the caravels on the "American" shores. While Nazi Germany annexed Austria almost without firing a shot, Zweig took the road of exile. After an English stopover, he moved to South America. Brazil, where he settled in 1941, was what the Italians would call the *ultima spiaggia*, a bittersweet expression that points to the last span of freedom before falling into nothingness. And this last span, terribly narrow, was quickly traversed. In Petrópolis, on February 22, 1942, he lay down after taking Veronal (Barbitone) and died; his wife Lotte lay next to him after having taken a lethal dose of morphine. The writer and his wife had decidedly nothing in common with the conquistadores. However, shortly before his death, Zweig conducted a brief study of the least conquering of all the conquistadores, the one whose name, slightly transformed, would return most often to lips and under pens over the centuries: Amerigo Vespucci, the man of "America." *Amerigo. Récit d'une erreur historique* appeared in 1942, four years after *Magellan*, which Zweig devoted to another navigator. There is something touching in this frantic quest to which the exiled man delivered himself, so near his death. He too was not the first to tackle the land of Brazil: there was Pedro Alvares Cabral, Magellan before him, and then Vespucci. In *Amerigo*, Zweig investigates a long-term "historic mistake": why did Vespucci give his name to the continent that Columbus had discovered a few years before him? His first name or . . . Vespucci was probably called Alberico, a more common name than Amerigo; he might even have become Emeric, if French had been the dominant language in the sixteenth century—Emeric of Vespucci Florentine, as Polo became Marc Paul Venetian. We find these unusual francizations in the eighteenth-century texts. But let us stop there. We will not return to this well-known case, of which the Florentine was the victim and not the craftsman. Moreover, it is possible that Vespucci died before learning that he had given his name to the land he had never claimed to discover. We will not return to the reasons that, one day in 1507, led a maniple of humanists gathered around Martin Waldseemüller, in Saint-Die, in the Vosges, to write the word *America* on the world map they introduced into their *Cosmographiae Introductio*. None of these scholars had certainly guessed that this map, elaborated with the inscription they had just affixed, would become the baptismal act of America. Besides what could this neologism, "America," be? It downsized to just be part of Brazil, its northern coast. We will not return to Waldseemüller's repentance, as he, in his new maps of 1513 and 1516, crossed out the word *America*, considering that Columbus, whose primacy he now recognized, had ultimately been right, and the *Mundus Novus* was nothing more than a portion of Asia. But America and its new name were already on the way to glory. As Zweig wrote, "never, or hardly, have chance and error developed such an audacious comedy."[51] Let us leave all that aside to return to the last lines of Zweig's book. For the exiled Austrian, Vespucci was an uncomplicated individual. Under no circumstances was he an *amplificator orbis terrarum*. He was not a liar, nor a fibber, nor a crook based on the reputation of others. He was what would later be called an honest man. That is why, according to Zweig, the name "America" has not been usurped. It immortalizes the man who first realized the novelty of the world unfolding before his eyes; it especially immortalizes a simple man, one of those self-made men that the northern part of the *Mundus Novus* would one day worship: "And maybe," Zweig speculates, "the name of an ordinary man like that, the name of a man belonging to the anonymous cohort of the brave, is better suited to a democratic country than that of a king or a *conquistador*. It is certainly more appropriate than if we had called America the West Indies or New England, or Spain or the New Land of Santa Cruz."[52]

The naming of America, the Americas, is the result of an accident, which has become grotesque by its way of being exceptional, as demonstrated by Zweig. The link between *America* and *Amerigo* is so surprising that it was long refuted. In *Le Nom de l'Amérique* (2006), Albert Ronsin proposed a beginning inventory (in French) of the variants scaffolded by a sometimes admirable ingenuity in the late nineteenth century.[53] For Jules Marcou, distinguished professor of geology who taught at Zurich and then at Harvard, *America* was the name that the original peoples of Nicaragua had given to a mountain chain that Columbus, Vespucci, and Pinzón had all seen and associated with gold. In 1890, Thomas Lambert de Saint-Bris published a scholarly essay under the programmatic title *Les Preuves de l'origine des noms d'Amérique de Christophe Colomb, ainsi que la croyance que l'Amérique fut la Colchide*. This time, *America* is derived from *Amaraca*, the name of the sacred territory of a Peruvian tribe. Thus Saint-Bris reactivated a legend that made America the land of the ancient Colchis. So Jason had discovered America before Abu Bakari II! So quibbling over who, between Columbus and Vespucci, was the first made no sense: at best, both were fighting the rear guard! They were mere plagiarists of the history belonging to another. But *America* could also be the contraction of *Ameracapana*, the name of a destroyed city in Venezuela. This is at least the opinion Alphonse

Pinart, learned linguist, expressed in an article in 1891. And so it continues, because this litany is at least as long as the list of the mythical owners of first places. As a codicil (the notarial terminology seems appropriate to me), I will add that the British did not remain immobile. For them, the eponym came from a certain Richard Meryk Ap, according to Richard Amerike, the Welsh patron of traveling who, in 1497, led John Cabot and his crew from Bristol to the coast of Newfoundland. To show his gratitude, the navigator placed the new geographical area under the patronage of Mr. Amerike.

Usually, things are simpler, more direct. The naming confirms the transition from open to manageable: the most apt toponym is sought to signify taking possession of the space and the reduction of its share of the unknown. Saints and kings are of great help in this votive effort. Having just arrived in what would one day be called the Bahamas, Columbus quickly set about converting the Arawak names into a string of more conventional names. The island of Guanahani, discovered on October 12, had turned into the island of San Salvador by October 14. Then Santa María de la Concepción arose, which later became Rum Cay, probably because a ship loaded with a cargo of rum was wrecked there. Drunkenness from geography meets fragility of the value system. Three or four days after the initial landing, Columbus turned away from the saints in the calendar to honor the Spanish royal family. An island near Santa María was dedicated to Fernando, King of Spain: it would be called Fernandina (the future Long Island). Another island whose native name he thought to be Saomete became Isabela (Crooked Island) in tribute to Queen Isabella. When the list of kings, queens, and saints was exhausted, the names tended to be redundant. Fernandina and Isabela are also noble etymons of some of the principal islands of the Galapagos. Unlike their Bahamian namesakes, they have retained their Spanish names. The archipelago of giant tortoises, later becoming Ecuadorian, remained Spanish-speaking, while in the eighteenth century, the Bahamas joined the British Crown. That explains the new naming of places. We know that Columbus was driven by ambivalent intentions. If on one hand he was forced to find new names to baptize the many islands he discovered almost every day, he also sought to adjust the geography of places to an older geography that piqued his imagination. He stubbornly sought the outpost of Asia, the island of Cipangu, today's Japan. On October 23, he headed to a large island, which his interpreters called Colba. It was Cuba. For Columbus, Cuba was Cipango, although on December 5, Cuba became for him Juana, in honor of Juan, the only male child of the royal couple. From Cipango to Quisay, the capital of the Great Khan, the path should not be long. However, it was not as short as he thought or wanted to believe. All in all, Columbus struggled in a triple onomastic indigenous system, Spanish and Catholic, incidentally

Far Eastern, the latter being largely dependent on Marco Polo and those who arranged their worldview based on the Venetian merchant's trip.

Later, the processes would diversify. Columbus had not thought to transform open areas into offshoots of Spanish territory. It is that, once again, he did not have access to the New World, but to the oldest of all, Asia. For his successors, it was more of the same. Since America was indeed a New World, it would be the double, or the avatar, of the land of the fathers. Logically, the European colonial powers that invested in overseas lands have strived to offer their citizens a reading grid to conquer space or spaces already conquered and "localized"—that is to say, spaces transformed into places wrapped in a reassuring onomastic. The most famous example is perhaps that of *Nieuwe Amsterdam*, a duplicate of the capital of the Dutch settlers, which later became *New York*, the reduplication of a city and a dukedom charged with history in the eyes of British applicants. It was both to mark possession and to reduce the strangeness of the unusual space. It was basically to "domesticate" otherness by a set of soothing metaphors. New York was named after the Duke of York, brother of King Charles II, who would erase from the Manhattan island's surface all memories of its inhabitants, the Lenape. At a time, New York was even New Angoulême, the name Verrazzano conferred to the site of the future metropolis he had discovered in 1524. There again, it was less the French city from the Charentes department that he honored than François I, King of France and Count of Angoulême. Examples abound. It was in all cases the first occupants of places who suffered from what Louis-Jean Calvet called a process of *glottophagy*.[54] If a cannibal had transformed the tongue of a conquistador into a delicacy after having cooked it according to the recipe of his choice, the tongue of the "Indians" had been massively ingested for its rapid digestion of language and identity. As Aesop claimed, the tongue is the best and the worst. As an organ or a system of expression, it is fragile.

The span of the "newness" of place was limited by seniority and the prestige of the European model that it duplicated. For Cortés, Mexico was none other than the New Spain. For this reason, it had the remarkable honor of being the double of the most prestigious of all countries: the one the conqueror came from or for whose benefit he performed a service (the distinction is significant, because without it, at the time of Columbus and Vespucci, the Italian maritime republics would have had their say from the other side of the Atlantic). Then there was a New England, which today still includes the same six states in the Northeast United States; a New France, which consisted of all the American colonies, from St. Lawrence to the delta of the Mississippi; a new Holland (Australia), and so on. Cities and regions were both doubled. Even the prefixes *nueva, nouvelle, new,* and *nieuw* were abandoned in favor of exact replicas of the European model. Where is Guadalajara? and Cordoba? In Spain? Yes. Mexico?

Also. And Paris? and Rome? In France and Italy, of course. But the City of Lights and the Eternal City are also in Texas, one in the north, the other in the south of the state bearing a Lone Star. And these two towns thrive among the other seven American Parises and four other American Romes. Europe has constantly left a breadcrumb trail to try to find—and to navigate—the maze of new spaces that open before it. But this thread, in a formula Gracq applied more modestly to the Nantes of his childhood, "takes on in its circumvolution the character of an irregular winding."[55] A final example, saving the best for last: Venice (Venezia), which gave its name to Venezuela. It is said that Vespucci made an association between Venice and the stilt houses of Lake Maracaibo that he contemplated from the bridge of the ship commanded by Alonso de Ojeda. Venice also gave its name to a Californian Venice. In 1900, the tobacco magnate Abbot Kiney, returning from a trip to Europe, decided to reproduce in all modesty next to Los Angeles the Italian original. He had canals dug and had palaces built. But he forgot the tides of the Pacific and the impertinence of nature. And the Barcelona traveler Xavier Moret commented, "What was, for example, the Grand Canal is now the Grand Boulevard. It is not the same, that is for sure, but there are fewer mosquitoes."[56]

Sometimes, the colonizer has kept the original names. Such was the case in Mexico. Meshico-Tenochtitlan was the Nahuatl name for the Aztec capital. Europe was no longer sufficient to fill the New World with names familiar to the ears of the new occupants. Repetition was circumstantial. Sometimes inspiration was drawn from the local reservoir to enrich the verbal constellations of the onomastic work. Gonzalo Fernández de Oviedo already suggested so, as did Francisco López de Gómara who, in his *Historia*, in chapter 208, explained with care the Nahuatl etymon of Guatemala—namely, *Cuauhtemallán*—"which means 'rotten tree' because *cuauh* is 'tree' and *temalí* means 'rot.' But it could also mean 'wooded place' because *temi*, which could be its component, is 'the place.'"[57] The first term was a good alternative. What López de Gómara poorly discerned was that the Aztecs and Mayans were themselves referring to very elaborate toponym codes. Men of the plateaus, the first used land elevations (*tepetl* or *tepec*) to name places and to symbolically represent them using pictograms. As for the latter, who lived in the plains, they relied on rivers, swamps, and so on to do the same. As the pictograms were too complicated to paint, they resorted to homonyms, a process presaging a decoupling between onomastics and the realities of the terrain. In Mixtec, as recalled Joyce Marcus,[58] feather (*yodzo*) had come to replace the valley (*yodso*, a spoken word with a different intonation). The effort expressed by the historiographers proves in passing that the Spaniards did not balk before the cumbersome task of learning the local language. In *Les Quatre parties du monde. Histoire d'une mondialisation* (2004), Serge Gruzinski, a specialist in the history of the conquest of the

Indies, defends, in a distanced approach, a vision that fights the static representations of the Spanish (and Portuguese) colonization of the sixteenth and seventeenth centuries. Animating a vast panorama simultaneously embracing Mexico, the Andes, the Philippines, or Brazil, Macao, and Angola, the Spaniards and the Portuguese had established a continuous flow between their possessions and their bridgeheads. To go to the Philippines or China, the Spaniards had long crossed Mexico and Central America to avoid the Portuguese—and, who knows, Adamastor—to reach the Cape of Good Hope. Well ahead of their time, their perception of the world had a global reach.

In this ample and fluid environment, some scholars, often religious, mastered a surprising number of languages. Martín de Rada, an Augustine Navarrese, traveled throughout Mexico starting in 1561. He then went to the Philippines, China, and finally Borneo. He died in the sinking of the ship that was to take him back to Manila. During his career as an evangelist, he learned the Otomi language of central Mexico, the language of the Cebu in the Philippines, and Chinese.[59] One can easily imagine he gracefully handled Spanish, Portuguese, Italian, and Latin. It is not unimaginable to think that he spoke some other languages that were more or less unusual. Sometimes it was men like him who helped name these places that their compatriots fashioned in their own image as many minor deities do. To accomplish this, Rada and the others based themselves on a perspective combining claimed idiosyncrasies and implied acculturation. In New England, the process took a slightly different turn: native names, although known, were immediately retracted. John Smith had admittedly collected the Algonquin names during his stay in Virginia, 1606–7, but upon his return to England, he made sure they were superseded by names more appropriate for a British ear. The original traces should disappear. Smith encouraged the future Charles I to work on the naming of places. The young prince of Wales did not have to be asked twice to use his imagination. The future Cape Cod became Cape James (named after his father), another cape became Cape Elizabeth (the name of his sister), and so on. Fortunately, his family was large. Later, Charles took up the task to which Columbus had spontaneously set himself prior. But nobody wondered what the "Indian" name of Cape James was, except John Smith, who did not care anyway. Meanwhile, the beautiful Pocahontas became Mrs. Rebecca Rolfe. A 1616 portrait represents her wearing the finery of a lady of the English court, an indefinable smile on her lips. Pocahontas was baptized in another religion, just like her homeland was.

At other times, places seemed to fiercely resist efforts of naming. Australia gave the impression of wanting to vehemently dodge any special improvised patronage. Was it that the newcomers cared less about local languages than the Spanish, once in their global empire? Yes, undoubtedly. Despite three centuries of hindsight and the teachings of the Enlightenment, it is possible that the

Aborigines had even worse press for the settlers in Australia in the nineteenth century than the Aztecs in New Spain in the sixteenth century. Barron Field (a friend of Wordsworth and Coleridge, the same who recommended to the reader to suspend disbelief when confronting fictional constructions), landed in the southern continent in 1825. He unsurprisingly ranted against the lack of awareness of the semantic landscape unfolding before him. He was more skeptical when facing the real than his friend was when facing fiction. He simply did not believe his eyes. The few names that had been designated up to that point had merely reproduced the Land Register of the British alma mater. To him, they seemed to be unfounded. The referent is disturbed in forced compartmentalization. And the peaceful village of Blackheath (literally "black moor"), located in New South Wales, attracted his wrath: "'Blackheath is a wretched misnomer. Not to mention its awful contrast to the beautiful place of that name in England, heath it is none. Black it may be when the shrubs are burnt, as they often are."[60] Never mind, the traveler's angry stay did not go unnoticed. A street in Sydney, Barron Field Drive, now bears his name. Sometimes, the streets are the obituaries of disoriented bad poets.

The great expeditions from the late fifteenth century and throughout the sixteenth century (we could extend the catalog until the eighteenth century at least, even to the twenty-first) have completed the establishment of a precarious equivalence: the opening of space was a prelude to the covering of place. And this covering was itself initiated by the colonial enterprise. A map is created; it seems free of referents, or at least it is supposed to be; the territory is seized, and it is reduced to the scale of the perceptible; next there is the "fluttering from one proper name to another, inside the frame that . . . is imposed,"[61] as Michel de Certeau said. We invent. And from cape to estuary, from forest to savannah, the army and the clergy cooperate. The holy water is poured into the sand of the beaches where the flags of the conquerors are planted. It evaporates, but the scent of incense persists, and the names do too. The baptism of places aims to deny the official existence of the first occupants of what was still space for Westerners. Reterritorialization takes place. One flutters, or even flies, robbing others of that which they are not supposed to have. The world is new. Or is it not true that what is new is empty? And then it is filled. Names proliferate; the words are printed on sheets of maps and the roundness of the globe. Does something remain beyond? It is astonishing. On his deathbed, the Tamerlane of Christopher Marlowe's 1587 play requests a map. One is unrolled for him, and he begins to list the inventory of conquests that led him from his native Scythia—he was born close to Samarkand—to Zanzibar and Persia. He definitely did not lack much to seize the last acres of the world. But he failed in his race against time: "And shall I die, and this unconquered?"[62] His confusion is great. Yet he knows what remained before him, just out of his reach: "My sons, see these tracts of land / Halfway between Cancer up to the West."[63] He then resolves to pass the torch to his heirs. Marlowe himself had an heir, the same as his Tamerlane: Sir Walter Raleigh,[64] the man who inquired about the Amazons there where Tamerlane had no chance of imagining them. The abstract line of the tropics became a reality. Space is being striated everywhere. The invention of place is about to be patented.

CHAPTER 5

The Measured Mastery of the World

Invisible Lines and Rings in the Water

One day, Bruce Chatwin, a great English writer and a big traveler, wandered into a street flooded by sunlight, in Alice Springs, in the heart of Australia. He noticed that people were going about in Land Cruisers, refraining from walking. He knew that the place was not suitable for pedestrians, because it served as access to the vast Australian desert. Chatwin went to Alice Springs to meet Arkady Volchok, an anthropologist who had initiated a survey of aboriginal sacred sites, and Volchok was to lead him into the bush. During his short life (he died at the age of 49), Chatwin had crossed the Patagonia and parts of Africa, Afghanistan and Nepal, but it was in the Australian emptiness that he discovered another way of reading space. According to the Aborigines, Volchok explained to him, totemic beings created the world, during Dreamtime, "singing out the name of everything that crossed their path—birds, animals, plants, rocks, waterholes—and so singing the world into existence."[1] Continuing his lesson, the anthropologist added that each ancestor had marked a trail of dreams composed of words and musical notes, a sacred wake that his descendants would endeavor to follow. "In theory at least," Chatwin said, "the whole of Australia could be read as a musical score. There was hardly a rock, or creek in the country that could not or had not been sung."[2] Indeed, the entire continent was virtually covered with song lines. The ultimate crime was therefore to sing off-key because "it un-created the Creation."[3] Uncreation is a constant threat to the balance that is just as precarious as the cultural and political environment of the Aborigines. In any case, it is an abolition, which itself is indispensable: that of the borders, these lines that materialize the barbed wires of the whites. For the Aborigines, moving around was vital: "The definition of a man's 'own country' was 'the place in which I do not have to ask'. Yet to feel 'at home' in that country depended on being able to leave it. Everyone hoped to have a least four 'ways out.'"[4] It was absolutely necessary to maintain good neighborly relations, because each place was nothing but an inflexion point in a life that intelligence and self-preservation rendered nomadic. These song lines shied away from the eyes. They solicited senses other than sight: touch, smell, and a magical hearing. The song lines were drawn before the eyes were crowned as the hierarchy of the senses (in the West); they were invisible and they were the guests of polysensorial excursus. It was about "escape lines" in the literal sense that Deleuze and Guattari have subscribed to this formula. Aborigines are not rooted in a territory or in a representation that is based on a conventional meter or a stable standard. They occupy a *threshold*; they illustrate the principle of transgressiveness, which is to occupy a permanent in-between, a third space pervaded of all the forces that arise and express the hinge between worlds. They draw a possible world that comes in the plural, while the White world, frozen in a representation that is ideally unique, declines in the singular. Polyvocal versus univocal. Pluriversal versus universal.

The mobility of spatial representation that Chatwin portrayed here is not an isolated case. As explained by the ethnologist Vishvajit Pandhya, the Ongee hunters of the Little Andaman, an island belonging to the Indian Archipelago located in the Gulf of Bengal, also forged a totally dynamic vision of their environment. Aware that they shared their environment with the animals they hunted and in turn with the spirits that haunted the island, they regarded the space as a moving frame associated with changing activities, so that, according to Pandhya, their maps did not reflect "the places in space but movements in space."[5] Space is the medium of nomad paths. No place can grasp nomadism in its entirety. I am convinced that an ethnologist would be able to cite as many similar examples. The Western view of the map is a variation among others. And as for lines—there is a multitude that would mimic the escape lines. Chatwin has evoked some that have inspired links to song lines among his interlocutors. Among them, he puts the *ley lines* that connected the old network of major megalithic sites, or the Sami "singing stones," which were also aligned. In bulk, he cites the "dragon lines" (telluric currents) of Feng Shui and the Nazca lines in the desert of central Peru, "a kind of totem card."[6] These examples are not isolated, but they are all foreign to modern Western culture, in which the idea of oscillation or rhizomatic mobility quickly becomes suspicious. However, the individual is not naturally sedentary; he is nomadic, or he was. The lines of the ways once followed are imprinted in his genetic heritage. Facing the new world that opened before him from the late fifteenth century, Western man could have resumed his original nature. During a parenthesis that was delightful, despite famine and distress, Cabeza de Vaca did. He wandered first beautiful and ragged, and then he espoused the escape lines of the First Peoples. But if man is not sedentary by nature he is certainly, in the playful lexicon of Deleuze and Guattari, a "segmentary animal."[7] It is possible to be other than sedentary, but it is not possible to be other than segmental. This logic cuts and separates, aims to identify, to assemble what is said to resemble at the expense of everything else; it separates the one who cultivates it into a circle or behind a line. It protects. Against what, against whom? Against the unknown, against the Other. Always *against*.

Jules Laforgue never frequented Australian Aborigines. If I'm not mistaken, his work has not attracted the attention of Gilles Deleuze. But it could have interested Georges Poulet, who for his part was more absorbed by the metamorphosis of circles than by the escape of lines. Indeed, in his *Moralités légendaires* (1885), Jules Laforgue portrays a Hamlet who, from the top of the diamond-shaped window of the Elsinore tower, made ripples "in the water: might as well say in the sky." And Laforgue added, "Such was the starting point of his meditations and his aberrations."[8] Seeing the water and escaping along the lines that go beyond the Sund and Denmark was a solution for Hamlet. But no, he preferred the round ripples of his spit, circular like the tower that protects him from an adverse world. Hamlet is in search of the perfect circle that delimits the intermediate zone between being and not being, somewhere between water and sky. Hamlet's ring is an additional round circle that Frank Lestringant failed to rank among the circular figures of the Middle Ages and the Renaissance, along with the sphere of the world, the king's crown and the tonsure of a monk.[9] The Prince of Jutland—was he an exceptional being, a being who is the exception to the rule? To pretend otherwise would be foolhardy. Moreover, we actually believe the myth. With the help of Shakespeare, the singularity ideal of an undecided hero is established. Yet the sailors who left familiar places in Europe to open up opposite spaces did little more than make circles in the wake of their caravels. "In the water: might as well say in the sky." They could ride the escape line *par excellence*, this horizon whose semantic emergence in the middle of the thirteenth century would have punished the liberation of desire and invited their spirit to soar. But this line definitely had an inborn defect: it does not comply with the axis of the navigation. Not truly port, nor frankly starboard, it did not accompany offshore movement. Rather it materialized the promise of an infinite forward momentum, a permanent beyond that is approached without ever being crossed. This line intimidated, crushed, and especially persistently thwarted any attempt at finitude. It evaded from the front. It evaded from the rear. High seas navigation made of the horizon a roaming circle that contained nothing and reassured no one. Columbus faced this challenging geometry, but Jason may not have: the Greeks stuck to the shores and kept landmarks in sight. But in *Paysage avec Argonautes*, which completes *Rivage à l'abandon* et *Médée-Matériau* (1982) and offers a splendid reading of the myth, Heiner Müller sums up in one line the in-between of Jason and of all sailors: "With the horizon, the memory of the coast faded."[10] Navigation in space is in the pure present, in the tension between an uncertain future and a memory that fades away.

Faced with the vastness of the world and the incommensurability of desire, there was nothing else to do but draw circles in the water, small circles of bitter disappointment that converted the navigators dispersed within the great maritime outdoors into the inheritors of the smallest of territories: the territory that confined the propelled saliva to the surface of the waves. This was still the starting point for some meditations and many aberrations. It was also an expression of the boundless hypocrisy of being. The navigator attributed to the sea the indecisive character that was the reflection of his own trial and error. *Character* is fitting, as revealed by its Greek etymology: the art of engraving furrows, of leaving an impression (*charassein*). For the Romans, *character* was the even markings made with a hot iron. The sea was deprived of the essential "character," and in front of the infinite line separating the sky from the sea, the sailor, for his part, could lack character. The sea remained for some time an elusive smooth space. We didn't have to wait for Deleuze to explain to us that modern man hates and fears the spaces through whose very smoothness he is opposed to anomic resistance. How can the sea and the new oceanic expanses be striated, when they are so rebellious? How can they be given *character*? The answer seems simple: it was necessary to tame the horizon line and its intolerable challenge through an appropriate symbolism. Owning the beyond. Conquering the horizon was as good as enclosing within the extended maritime limits; it was to make of the infinite something undefined. But the indefinite is potentially open to definition. It is intended to draw back with time and the generation of new experiences.

The Measured Mastery of the World

Deleuze and Guattari are the initiators of escape lines that are as invisible as the song lines of the Australian desert. For his part, Carl Schmitt had quite a different conception of what the line could represent. We know Schmitt's career path. Compromised by the Nazi regime, he was imprisoned between autumn 1945 and spring 1947, without ever being reintegrated into the German University. Schmitt had understood what the link between *Ortung* (location) and *Ordnung* (order) meant. Had he fully understood what the ultimate consequences were of a too-close association between such overindulging paronyms? It is that it does not necessarily lead to the peaceful coexistence of peoples. Moreover, the regime that he had supported had given many examples of coercive organizing of place whose effects were often tragic. For Schmitt, the measure of order is the *nomos*, which establishes the configuration of the territories and the political, social, and religious balance. *Nomos* was the Greeks' doing. As Schmitt pointed out in *Le Nomos de la terre* (1950–88), his key work, Plato had already mentioned it in his *The Statesman* (*294b*). But it is well in the age of the great discoveries that *nomos* took on a new dimension. It tended to correspond to an "international spatial order of the Earth as a whole."[11] Its pivot was obviously Europe, the self-proclaimed *omphalos*, and its basis a Eurocentrism that made sense, even if the inclusion of otherness began to germinate in some fine-tuned minds. It is thus how the sketch of "thought by global lines"[12] (*ein global Linien-denken*) came about, which would become characteristic of Western ideology of modernity or the common point among various Western ideologies of modernity, as long as one prefers the plural to the singular. Deleuze would have rather written that one now aspired to impose a "*métrise*" of the world and that the overcoding machine had been set in motion, whose aim was to stop the famous escape lines according to a logic of power. This criticism would have hardly attracted Schmitt, a man of the established order and of the repressive apparatus of the state (they were simply legal).

Near Valladolid, on the banks of the Duero, Tordesillas is a small town where heat waves rage in the summer—a bit like Alice Springs, some distance away. But in June 1494, a certain agitation reigned there. Emissaries of Queen Isabella and John II of Portugal gathered in a small palace, which is now known as *Casa del Tratado*. John II had not appreciated the Papal bull *Inter cetera divinae*, issued the previous year by Pope Alexander VI. It had at least two flaws in his eyes: that of being Spanish and that of having emanated from a Borgia. Here are a couple of good reasons that revoked in doubt the validity of the arbitration, knowing that the lands discovered west of a meridian one hundred leagues from the Cape Verde archipelago belonged to the Castilians, while those that were deployed to the east of this line belonged to the Lusitains, except that land that already had a (European) owner. At Tordesillas, an agreement was reached: the line was moved 370 leagues west of Cape Verde. The case turned out not too bad for the Portuguese because this correction then allowed them to settle in Brazil. But in 1494, the Brazilian coast, provided that it had not been trampled on by Abu Bakari II's rowers, was still protected from ultraocean invasions. Cartographic abstraction had just triumphed, a few decades after perspective lines appeared in Western European painting. Erected in the middle of nowhere or nearly nowhere, Tordesillas became the scene of a momentous decision: a world that was still *terra nullius*, land belonging to no one, had been divided up. Because, it is necessary to repeat, water was worth less than land, and the indigenous peoples met on these lands counted for even less than nothing. Later, the question would be seriously posed if they even had a soul.

In 1524, the two contenders for control of the world—the others being relegated by them to the rank of pirates—began a new controversy. A line was to be drawn in the Pacific equivalent to the one that the Treaty of Tordesillas had helped put into place in the Atlantic. Therefore, there was a meeting in Badajoz, in Extremadura, near the Portuguese border. Francisco López de Gómara told the story of this negotiation in his *Historia General de Las Indias*. During two months, globes had been contemplated, maps perused, travel logs dissected. But nothing helped: the lure of spices in the Moluccas was so strong that not an inch would be given. The line remained even more abstract than ever; it remained to be traced out. The negotiators went into town to cool off and take a break, which would not be easy, as it was May and it was already getting hot. At a street corner, a kid who watched the clothes his mother had hung out to dry asked the experts "if it were really they who divided the world in the company of the Emperor."[13] And as they answered in the affirmative, "he lifted his shirt and showed his buttocks and said: 'So have the line (*raya*) cross here, in the middle.'"[14] And the delegation burst out laughing. The Badajoz rascal had demonstrated having a better sense, safer than the theories of the learned geographers, sailors, and well-considered prestigious diplomats. Is it he who pointed out the ridiculousness of their approach? Is it he who made the confabulations fail? It was not until 1529 that a treaty would finally be signed in Zaragoza: a line would now cut through the waves 297½ miles to the east of the Moluccas. As before for Cape Verde, it was necessary to determine specific landmarks within a blurry geographical environment. The problem remained unsolved, and in uncertainty, a margin of leeway remained. A few kilograms of additional spices would have tipped any trade balance to the right side.

The horizon, at the end of maritime space and that of land as well, was reduced to an imaginary line to be reproduced at will on maps. The indefinite was reduced to human scale by a process of accomplished abstraction. Schmitt inventoried some of these lines, starting with the *rayas*, whose path depended on a legal agreement, a common *ordo*. Then there were the amity lines, "lines of friendship," which implied enmity lines. Their emergence accompanied that of the Reformation and Counter-Reformation after the Cateau-Cambrésis peace (1559). It also accompanied the entrance on the oceanic scene of France and Great Britain in addition to Spain, while Portugal fell the day after the battle of Alcacer-Quibir, in 1578. Meanwhile, the meridians forced themselves on the large geographical theater. To the east of the prime meridian, the Cateau-Cambrésis rules were fully applied; to the west, as an American political scientist wrote, "might should make right, and violence done by either party to the other should not be regarded as in contravention of treaties."[15] There were friends to the East, but what lied to the West was neither seen nor known. Pragmatic, arbitrary, cynical . . . The appropriate qualifier may be chosen. The amity lines were based on a new method of dividing up the globe, which included a Meridian calculation. So by where did the meridian of reference pass? Not yet by Greenwich. For Richelieu, no meridian should pass beyond the island of El

Hierro (Iron Island). In 1634, a major geopolitical battle raged. Its main stake was determining a longitude of reference that would allow for better coordination of new discoveries and the dividing lines. The cardinal then decided that the island of El Hierro, which is the westernmost and smallest of the Canary Islands, would house the meridian that cuts the world in two: the Old to the east, and the New to the west of the imaginary line. The Canary island, according to Richelieu's mathematicians, had the advantage of being exactly at 20° to the west of Paris. El Hierro became *La Isla del Meridiano* and the Canaries no longer represented the end of the world: they adjoined the center of the world and the ultimate indicator of the Old World. Thereby, Richelieu opposed those who preferred placing the line via the Azores. But it did not prevent the Dutch from choosing yet another marker, not so far away: the summit of Teide, the volcano that dominates Tenerife.

At their peak, 14 zero meridian competitors can be identified. We are lost. The treasures have been misplaced. In *Le Trésor de Rackham le Rouge* (1944), Tintin and his friends are searching at the bottom of the sea for the galleon commandeered by Knight Haddock, Captain Haddock's ancestor. In the middle of the ocean, they fail. They have, however, the exact coordinates of the wreck. But the reference meridian is not that of Greenwich, or even that of Timbuktu, which the captain gruffly pushes aside. Tintin has an answer to everything: "Knight Haddock, for his part, had certainly counted by taking into account the Paris meridian as the meridian of origin." And his descendant, delighted by the sagacity of his friend, exclaims, "Billions of blistering blue barnacles! You're right! How did we not think of it sooner?"[16] Anyway, we never think about it soon enough. Indeed, it was long impossible to accurately calculate longitude. "'Discovering longitude' became synonymous with 'trying the impossible,'"[17] says Dana Sobel, who has retraced the history of calculating in *Longitude* (1996), a small popular book that became a bestseller because obviously the subject interests many people. Dana Sobel recalls the efforts of Galileo, compiling a table of ephemerides based on the eclipses of Jupiter's moons but recognizing all the while that the mere heartbeat of the astronomer could distort the measurement. Galileo was certainly less eccentric and cruel than Sir Kenelm Digby. It is said that he invented the wine bottle, the shape of which we all know. Regarding the calculation of longitude, he had imagined placing aboard the vessels of her gracious majesty an injured dog that was expected to bark when his bandage, left back on the shore, was soaked in a miraculous solution of Powder of Sympathy at the exact moment when the sun struck the London meridian. It was not until the invention of John Harrison's chronometer, in the eighteenth century, that the problem was finally solved.

"In keeping with Iron Island's ancient prerogative, the susceptibility of each nation was formed, and the maps kept a desired uniformity,"[18]noted in 1823 the naturalist René-Primavère Lesson, who feared that every nation attempted to make the meridian pass by its own capital. In 1884, the issue did not arise in the same terms. The zero meridian was officially residing atop a grassy knoll, in Greenwich, on the territory of the main colonial power of the time. In Paris, they kept a local time for a while, but it was delayed by nine minutes and 21 seconds. A discrepancy with Greenwich Observatory was necessary, as Tintin had noticed. In *Le Méridien de Greenwich*, which I have already quoted, Jean Echenoz depicts the illegal activities of a maniple of spies and mercenaries on an island lost in Oceania, whose only peculiarity is that it is crossed by a magic line that indicates a "lean and high stele of gray concrete, erected in the middle of a horde of barbarians bushes."[19] Despite what is announced by the title of the novel, this line is not the Greenwich meridian line but the line of date change, otherwise called the meridian 180, the antemeridian. As for the island of Echenoz's heroes (or antiheroes), it is in the same ocean as the Island of the Day Before,[20] which Roberto de la Grive (Umberto Eco's character), the last survivor of a ship run aground a cable's length from the shore, wondered how to reach, since he doesn't know how to swim and building a raft was unsuccessful. Would he die of hunger and thirst the *next day*, such a short distance from the island that he saw the day before in the coralline waters of the Solomon Islands? Byron Caine, the instigator of the (false) project that ignited the lust of Echenoz's spies, would certainly have agreed with Roberto de la Grive: "This is a twisted meridian . . . twisted and swimming. It slips into the water from one pole to the other, without passing through any land. I guess it would be difficult to live in a country where the day before and the day after would only be a few centimeters apart, one could get lost in both space and the calendar, it would be unbearable . . . A wall could have been built to divide the island into two dates."[21] Caine makes the most of the situation. He asks for two days off per week, considering that there are two Sundays on the island. And this additional rest makes him a philosopher: "That we have to divide the world by this makeshift line, is the proof that we never got to reconcile time and space, to combine them together."[22]

The third global line indicated by Schmitt is more recent. It dates from the nineteenth century and sanctioned a new divide between the hemispheres: instead of the traditional north-south partition, an east-west divide, or rather, a west-to-east divide was established. This initiative was the response of the New World against the Old, and its foundation was the Monroe Doctrine conceived in 1823, designed to protect the special interests of the United States of America. The delineation of this zone occurs at 20° longitude west from the Greenwich Meridian and at 180° longitude in the Pacific (at the date change line). We note that no line is designated to protect the interests of the south toward the north; it is declining in its western or eastern version. Incidentally, Schmitt forgot a fourth dividing line. It is the line that separates the goal line from the rest of the football field to form the penalty area. Is it worth mentioning? This line is regularly the talk of crowds of spectators on the lawns of stadiums, and it is no doubt the most striated of territories. Osvaldo Soriano, who knew the full measure in the matter of escape lines, made a few comments in the short story "The Last Days of a Happy Goalkeeper. A Century after the Invention of the Penalty."[23] This line was invented after a match on September 15, 1891, which opposed Notts County, one of the two Nottingham teams, and Stoke City. The game caused many reactions. On the goal line, a player from Stoke City indeed blocked a ball with his hand that would have allowed Notts County to even the score during the very last moments of the match. His lack of fair play was evident, but it was impossible to give the other team the goal. The referee then opted for the only compensation that the incomplete rules offered him: a direct free kick from thirty centimeters off the goal line. Obviously the entire Stoke team occupied the ridiculous gap between the line and the striker, who had no other recourse but to catapult the ball against a solid forest of guards. The indignation was intense. The penalty area was invented seven years after the consecration of the Greenwich meridian. A seven-year reflection that, however, did not change the Monroe Doctrine.

Back to Greenwich and, why not, to the exotic vicinity of Midway. Byron Caine was right: too much makeshift spatiotemporality eventually destroys the delicate balance of the world and its representations. But the fault does not lie with the designers of the modern atlas. Nor with Voltaire. Even less so Pascal, who said that "three degrees of latitude reverse all jurisprudence; a meridian decides the truth. Fundamental laws change after a few years of possession; right has its epochs; the entry of Saturn into the Lion marks to us the origin of such and such a crime. A strange justice that is bounded by a river! Truth on this side of the Pyrenees, error on the other side."[24] Whenever there is a fault, it returns, as was so well understood by Pascal, to all those who cast the world into a reading system that governs the abstract reality of things, knowing that among these "things" are stored the inhabitants of distant lands. The representation presides over a reification of certain portions of the real yet abstract when they are not merely putative. Geometry rediscovered Greek theorems in the sixteenth century: Euclid's *Elements* were translated into English in 1570, for example. The association between geometry and geography, which does not date from yesterday (e.g., medieval rhumb lines), would nevertheless be reinforced. Geometry contributed to reducing geography to a discipline that establishes the scale of the perceptible while setting the framework within that which exercised the political "*métrise*" of the world, the rampant oversimplification of the Other and of "his" spaces. Of course, this possessive adjective was free from any legal or ontological validity under the eager gaze of the European conqueror.

The world, at first, remained unknown in most of its parts, whose very existence was unthinkable, or at least inconceivable. Then we began to suspect the existence of a whole that certainly resisted the influence of practical experience but that intellectually had become apprehensible—in every sense of the word, somewhere between "*métrise*" and terror. Theories on the existence of an antipodal space had been dusted off—the theoretical projection of Australia that had not yet been approached by European navigators. Utopias played a leading role, and the most famous of them, *Utopia* (1516) by Thomas More, which according to Paul Zumthor, would close the gap "in order to organize by and in the text: or rather, the narrative creates it, a space of representation in which perhaps the vertigo of what remains to be done to survive in a world that has lost its measurement is resorbed."[25]

But the link between the supposed real and its fantasy representation was mainly woven by maps, growing ever more popular and more and more numerous, all in accordance with a hesitant reality or with a pure desire. The world was going to be perceived as a comprehensive place, accessible, controllable, and enclosable. We then enter a new mental regime that Brian Harley, geographer and author of several founding essays on mapping, has baptized as mapmindedness,[26] which roughly corresponds to a "penchant for cartography," of which Mercator's *Atlas* is emblematic. According to Frank Lestringant, this move from a vision of the world in which reigns "non-stop short-circuits between languages, images and distinct knowledge" and "the disconcerting makeshifting," which consists of bringing together "the remains of an ancient manhandled, battled knowledge, mixed together with the most unpredictable innovations and the most insolent naivetés"[27] corresponds to the end of the age of cosmographers, in which André Thevet, the subject to which his essay is devoted, was yet fully invested. The shortcut would hardly be rapid in affirming that this transition also marks the end of the Renaissance and the beginning of the modern age. But the (European) modern age equates itself with the assumed entrance into the era of colonialism. "Modernity," writes Walter D. Mignolo "is the name of the historical process in which Europe began its progress toward world hegemony. It carries a darker side, colonialism."[28] Mignolo adds, "There is no modernity without coloniality, because coloniality is constitutive of modernity."[29] We can see that cartography was the geo-, topo-, iconographical tool of this systematic colonial intrusion.

Cartographic Imposture: One's Self in Lieu of the Other

In 1572, Mercator chose to place his sketch of the world under the aegis of Atlas. But it was not the Titan who supported the terraqueous globe on his shoulders. Atlas became a mathematician and astronomer, a philosopher at times, who supported the earth with his calculations rather than his brute force. Geometry and the sciences then consolidated a career promising a brilliant acceleration throughout the arc of modernity. But this updated map reveals a disturbing truth: geometry is never dedicated to pure abstraction when applied. It serves as a prelude to a tangible representation of space. According to Ptolemy, the map is organized from a political, economic, and cultural landmark that radiates to the peripheries it designates as a self-proclaimed hierarchical system. We have seen it: the *Omphalos* complex activates whenever the layout of the world is concerned, whenever an extension is in progress. The materialization of local representations is exercised at the expense of others, because few spaces are completely empty. The degree of occupation of a place depends on the density of its population, but to the conqueror, it is mainly dependent on a fleeting impression, an oblivious glance. Even the desert, despite the stereotypical view of a tenacious orthodoxy, is populated, albeit more sparsely than other areas. It is often, if not always, that the deliberate blindness of the conqueror and his assumed bad faith produce a distortion in the point of view, which the map reflects or amplifies. The latter straightforwardly configures the application of geometry to a vision of the world that is interested, and therefore interesting, to the observer. The map is the instrument of domestication of the territory of the Other, who himself undergoes a subtle but inexorable Other-ing.

Allegory has remained on the margins of geometry, openly at first, then in a more implicit way. Sometimes natives are represented in a grotesque manner: the anthropological allegory then seemed doomed to an immediate political stake—namely, diminishing its target, minimizing its diameter. The Other was blended into the bestiary that the cartographer attributed to places; and the other was reproduced in as fanciful a fashion as the *tapiroussou*, the half-bovine, half-donkey creature that Jean de Lery believed he discerned in the land of Brazil and that was none other than the tapir. Sometimes, an artificial reduction in the gap was created by dressing (or redressing) the Other "as European," and his (usurped) territory was framed by the border lines that certainly recalled the principle of the boundaries separating the European states but that did not correspond to any concrete reality. This charade was all the more striking because no territorial precedence was given to the local First Peoples. Before the conquest, their environment was still amorphous. It was up to the colonizer, an almost divine figure, to shape the space, to take some clay and shape the place in his own image. Closer to home, further from Eden, *fingere* meant "to knead." The invention of place is a real fiction, a kneading or working over of the real. As for fiction as we know it today, it is itself a place, the "mixed field of production and of illusion,"[30] as Michel de Certeau says. A land for which the page is a map. Or a page whose land is the map.

During the Renaissance, the map quit trying to stage historical depth. It broke with medieval mappemondes existing prior to the great voyages of discovery. Now it wanted to instantaneously capture a fantasized space that was to be transformed into place. It was a pure surface, the symbol of a white space, a sure sign of virginity that was intended to yield. In short, it was subject to the *droit de seigneur* assumed by its "inventor." It was for the Western map as with the master's paintings: only occasionally did it incorporate an explicit diachronic. In the art world, one of these exceptions is a painting by Nicolas Poussin, *Paysage orageux avec Pyrame et Thisbé* (1651). Using a *mise en abyme*, the artist represents in two different planes of the same landscape the successive suicides of the lovers about whom Ovid wrote in the *Metamorphoses*. The use of this method is not so common in painting. It is even rarer in mapping since the late Middle Ages and early Renaissance. As an example, some of the maps that the humanist Hartmann Schedel recorded in *La Chronique de Nuremberg* (1493) can be cited. Schedel represented in synchrony the stages of the history of the world, but his work mobilized history from an exclusively Christian angle and only evoked the scenes of a mythical past because they announced an assumed future that conformed to an established model. In Europe, place is inevitably part of a moment that announces a possessive future. In fact, this mind-set does not refer to any cultural universal. There are many maps that locate places in the duration rather than in the moment. The Aztec *lienzos* clearly demonstrate that, rather than geometry, they privilege the relationship between space and the seminal event. As Elizabeth Hill Boone emphasizes, "one of the dominant features of [Aztec] mapping stories lies in the fact that they present the beginning and end of a story within a single visual composition."[31] The typical story is that of the quadricentenary migration from Aztlan, the mythical island of origin, "place of whiteness" (in Nahuatl) to Tenochtitlan, the island city that Cortés's men will seize to make into Mexico. Note, however, that these maps evoke their own history and not the history of others, such as the medieval mappemondes.

Most often, natives were erased from European maps. The cartographer pretended that the colonizer had seized uninhabited virgin land. But the voluntary retractions also involved antagonist religions. It happened that rivals' sacred places—such as those of Islam in Europe or, later, those of Orthodox Religion in the Soviet Union—were "forgotten." Moreover, even within nations, a whole social class or ethnic group might undergo the same procedure. The obliterations occur whenever several competing representations vie for a single space, which is subject to the desire of one or more pretenders. In one of the studies assembled after his death in *The New Nature of Maps* (2001), Brian Harley analyzed the phenomenon of omission through the example of the mapping of the New World. If it was the "new" world, it is because everything that had been part of the "old" world, or otherwise said the previous world, had been carefully erased: "The same maps were also the subliminal charters of colonial legitimation. As is the case of maps of the English landscape, English maps of the New World exercised power through the categories of their omissions. These silences applied especially to Indian civilization. We ask ourselves, where are the traces of Indian occupation on the land?"[32] There were some exceptions, such as the map of the Hudson Bay Company, but if it mentioned the indigenous peoples, it was for commercial reasons and not for charitable purposes. To illustrate his point, Harley evoked the engraving that accompanied the map of the famous Hermann Moll, who lived between the second half of the seventeenth century and the first two decades of the next century (we never knew what country he was born in, which is surely a way to annoy a cartographer). In the background of the engraving are the Niagara Falls. In the foreground, there is a large population of beavers . . . but as for indigenous peoples, there are none. Humanity is eclipsed in favor of wildlife, whose merit is both to signify an exoticism crossbred with the mythical and to justify the monopolizing. The trapper remains in the minds a better agent of colonization than the soldier because he is peaceful; he *naturalizes* the plundering at work.[33] Indeed, Harley adds, "The maps have to be seen as part of a wider colonial discourse, one that helped to render Indian peoples invisible in their own land. Cartographers contrived to promote a durable myth of an empty frontier."[34] This myth was perpetuated until the twentieth century. Harley recalled an informed comparison made by Karen Ordahl Kupperman, an American geographer, according to whom America was seen as a *tableau vivant* in the economy in which the arrival of the Europeans meant the rising of the stage curtain and the beginning of the play.[35]

Cartography kept quiet about its palimpsest nature. All creation is substitution; it conceals a loss or an erasure. As Kathy Prendergast, we could paint a map of North America reproducing all toponyms including Lost, which is also the title of one of her paintings from 1999. The map would transform itself into a real palimpsest if it combined the successive seizures of a same space. It would then acquire this historical depth that the cartographer denies it. Obviously, there are many who talk about that which is no longer. But usually, the erasures go unnoticed, because what is apparent is more visible than what is not. This truism seems silly—the "seems silly" is a truism in its own—and, a priori, there is no need to be a confirmed phenomenologist to agree, but its application in practice constitutes a major and constant political stake. One thinks of the dissimilarities and the mutual obliterations that characterize the Palestinian and Israeli mapping surveys in the Middle East. The space is transformed into a controlled place within a system that emphasizes certain aspects in order to omit others. To interpret a map, to interpret the seizing of place in an arbitrary (subjective) standard is also to flush out the silences and retractions. This is what Harley called the hidden agenda of the map. Any message can be read between the lines.

While the maps of the New World included many blank spaces—"space cracks,"[36] as nicely put by Frank Lestringant—maps of Africa, according to a seemingly apparent paradox, long remained gleaming and/or covered with text. It was not until the nineteenth century that they integrated the white spots that were soon to be colored. The difference in treatment is due to the forced muteness of the classical culture with regard to the New World, whereas Africa was the subject of old, even ancient stories (on the sources of the Nile, the legend of Prester John, etc.).[37] So the African maps were filled, based on myth, for lack of having a direct knowledge of the terrain. But there is something else. This evolution reflected a fundamental change. The white spots were substituted for the story even though (geographic) science felt capable of absorbing, of imposing a minimal toponymy and an "objective" reality at the expense of a textual, visual, and sometimes pictorial representation. The recognition of the void coincided therefore with the conviction that the filling in of the map was within the cartographer's reach. And, as luck would have it, the term *cartography* began to spread in 1850, in France, under the leadership of Manuel Francisco de Barros e Sousa, Viscount of Santarém.[38] As for the white blanks on maps, we owe them to M. Jean-Baptiste Bourguignon d'Anville. The invention does not go beyond the eighteenth century. As noted by Peter Turchi, "thanks to Jean-Baptiste Bourguignon d'Anville and his colleagues, a blank on a map became a symbol of rigorous standards; the presence of absences lent authority to all on the map that was unblank."[39] Mapping definitely deserves for its history to be told, because this story speaks volumes about the relationship between reality and fiction.

Against Graphocracy, for a Different Cartography

The impostor stayed a long while. It is against the arbitrary filling in of the white spaces by the whites and against the oblivious practice of the Other, the victim of a "cultural genocide" in the words of Brian Harley,[40] that the writers of the postcolonial era were established. But they also rose up against the delirious categorization of colonialism at the height of its quest for legitimacy. For a long time, the Other was arbitrarily designated to be transformed into a freak show and has been made light of. It was necessary to reject projects such as Clark's *Chart of the World* (1822), which is cited by Ute Schneider, that divided the people of the world into savages, barbarians, half-civilized, civilized, and enlightened on the basis of a symbolic color code ranging from pink for the savages to yellow for the "enlightened." In *Running in the Family* (1982), Michael Ondaatje, a Canadian writer of Sri Lankan origin, returned to the traditional maps of his native island. The narrator of the novel, autobiographical or autofictional, refers to cartographic "speculations,"[41] establishing an equivalence between myth and "eventual exactness"[42] that would result in this way: "Amoeba, then stout rectangle, and then the island as we know it now, a pendant off the ear of India. Around it, a blue-combed ocean busy with dolphin and sea-horse, cherub and compass. Ceylon floats on the Indian Ocean and holds its naive mountains, drawings of cassowary and boar who leap without perspective across imagined 'desertum' and plain."[43] The desert is definitely the typical scene where the "out of place" character, the lawful occupant, the undesirable one, operates. Ondaatje adds, "At the edge of the maps the scrolled mantling depicts ferocious slipper-footed elephants, a white queen offering a necklace to natives who carry tusks and a conch, a Moorish king who stands amidst the power of books and armour. On the south-west corner of some charts are satyrs, hoof deep in foam, listening to the sound of the island, their tails writhing in the waves."[44] I do not know to which maps Michael Ondaatje refers, but this representation of Ceylon is the umpteenth version of a paradigm that has for centuries governed what Brian Harley defined as "subliminal geometry"[45] of lands to conquer. Sri Lanka is Ceylon, which was Taprobane.

The French legislature was rather uneasy when, on February 23, 2005, it passed a law that was more than controversial; the legacies of colonialism, if they are many, are never positive.[46] Indeed, the obligation of others to undergo a colonial apparatus is by its essence *negative*, in that it signifies the partial *negation* of the specificity of the Other. Among the many charges that resulted from Western control over a part of the world, we can include a representation of spaces, perfectly ethnocentric, that the geographic maps have sanctioned and relayed. Just when the process of decolonization was initiated, the revocation of the correctness of this monofocal vision of human geography was undertaken. The writers did not hesitate to take up the cartographic thematic in order to highlight any discrepancies from a point of view that it was no longer acceptable to share.[47] In the vast reservoir of postcolonial literature, we can select almost at random three novels that would enable us to hone the point of view, or rather, the idea that an advantageous plurifocality could prevail. There is of course *Running in the Family* (1982) by Michael Ondaatje, but also *Kartography* (2005) by Kamila Shamsie, a Pakistani author. As for Nuruddin Farah, he published *Maps*, translated into French under the title *Territories*, in 1986. It was about the importance of cartographic representations in one of the most troubled regions of the world: Somalia, his country of origin.

In Ondaatje, the narrator (the author's twin), temporarily leaves Toronto to go to Colombo. Before leaving, he contemplates some maps tacked to the wall in his brother's room. The display inspires more thoughts on the *Tabula Asiae* and the representations of Sri Lanka formerly presented by Western navigators.

In *Territories*, Nuruddin Farah gives great importance to maps, because they indicate the missing homeland of his hero, the young Askar, born in Ogaden, a region of eastern Ethiopia that for others corresponds to western Somalia. Passionate about cartography from an early age, Askar finds in his uncle Hilaal (a university scholar in Mogadishu) the ideal partner. Askar's passion is shared by Karim, Kamila Shamsie's student in *Kartographie*, whose ambition is to draw the map of Karachi, to name some of its still anonymous streets, to even the score with his girlfriend Raheen, who denounced in his project an infringement of the genius of place. Karim and Raheen like to indulge in the game of anagrams. One of them is particularly successful: the cartographer becomes a "graphocrate"[48] and mapping a graphocracy.

Nuruddin Farah lists several examples of the graphocratic power of mapping. Hilaal informs Askar that "Africa, in Kremer's map, is smaller than Greenland. These maps, which bear in mind the European's prejudices, are the maps we used at school when I was young, and, I am afraid to say, are still being reprinted year after year and used in schools in Africa. Arno Peters's map, drawn four hundred years later, gives more accurate proportions of the continents: Europe is smaller, Africa is larger."[49] Kremer is the common name of Mercator. The maps produced by the projection system that he developed are indeed questionable. Their mistakes all contribute to constructing the same isotopy: the elements of the North surpass in size those of the South, which is often wrong. Before the Berlin Wall fell, the Soviet Union (22.4 million km^2) seemed larger than Africa (30 million km^2), and the area of Scandinavia seemed equivalent to that of the Indian subcontinent, which is exactly three times as large. These proportions, now corrected on most globes, were still in force in the seventies. Dating from 1974, Arno Peters's map modified the Mercator projection by shifting the world in relationship to Europe and restoring fairer proportions.

This map created a huge controversy at the time of its publication. Peters was accused of deliberately borrowing Third World positions. But much earlier, the German mathematician Johannes Heinrich Lambert (in 1722) and the Scottish clergyman James Gall (in 1855) had attempted to correct the distortions in the Mercator projection.[50]

In 1995, within the context of a California exhibition that was dedicated to him, the Argentine artist Guillermo Kuitca had commented on several of his paintings reproducing maps of cities and countries scattered around the world: "'I don't want to make a geographic or geopolitical statement,' Kuitca says. 'It's a map; it doesn't matter so much whether the map is of Mexico or Norway. These maps all look alike, but they are of places that don't really look alike. They only look very much alike when you present them as a map. In a way, they're just names, not the real places.'"[51] Born in 1961, Kuitca has lived at the confluence of two diasporas: the one that marked the fate of his grandparents, Russian Jews who fled to Buenos Aires under the Nazis, and that of his own generation (victim of the Videla and other junta), which has often been forced to take the path of exile. The extreme mobility characterizing Kuitca's family history partly explains his passion for pictorial maps. But the impact of the map seems deeper than he says. The map certainly dissociates from reality, but it configures reality as well. The image that we have of places is modeled, in general, on the one that is conveyed by atlases, which are themselves increasingly electronic. We are returning to the misleading evidence that Arno Peters denounced.

Graphocracy, which resulted in the baptismal frenzy of the colonizer, caused serious trouble for the new states emerging from decolonization: Pakistan, as well as Sri Lanka and Somalia. What should have been done? Resign themselves to heterogeneous onomastics, which are imposed and endured? This solution would have been much easier since Western maps actually permeated minds. To Karim, Raheen contends that the mapping has no future "because all the maps have been made already, right? What are you going to do? Discover a new continent and map it?"[52] What Raheen forgets is that the map is not an absolute that establishes once and for all the geographical truth. All maps shift and let show through, sometimes by filigree, that which they strove to hide: the Other, and the precedence of his presence. All is not lost. As with Kathy Prendergast, what is lost displays a discreet presence. In *Maps*, Nuruddin Farah tackles the subject from a different angle. The academic Hilaal reviews the names of the African states. Nigeria owes its name to Flora Shaw, the mistress of the administrator Lugard. The name *Zaire*, which Mobutu Sese Seko, imbued with Africanization, claimed as native, actually derives from Portuguese.[53] Hilaal is able to establish what would be the etymological source of the names he recites. But when the memory of a distant past dissipates, the game becomes more complicated. Gambia is named after the river that runs through it and, in local languages, before British colonization, it was called Kamby. It is believed, however, that the place name derived from *cambio* because, as early as the fifteenth century, the Portuguese had set up trade stands along the river and organized commercial exchanges (*cambios*), a prelude to the slave trade. There are still some remains of these stands in the bush today. Alex Haley tells the story in *Roots* (1976). To my knowledge, there has never been an attempt to reclaim or return to the onomastic sources in Gambia, although the poet Tijan Sallah, in *Dreams of Dusty Roads* (1993) denounced the conservation of street names honoring, among others, Lord Wellington or the historian Buckle, "names which are neither local nor familiar."[54] As for etymons of Ethiopia (via Greek) and the former Abyssinia and Sudan (via Arabic), they all refer to the skin color of their inhabitants. Somalia, according to Hilaal is "unique. It owes its name to the Somalis."[55] In the postcolonial discourse, sometimes place names, depending on whether or not they are a mark of European occupation, become a pretext for a discourse whose matrix is nationalist. The identity of the One will always be considered more solid and therefore more legitimate than that of the Other, especially when the Other is the neighbor of the One. Together with the principle of territorial precedence—that mapping underpins since it tends to erase the strata of prior presence—the principle of legitimacy is part of a claim to universal significance, which is the cause of many conflicts. In short, the map could become the perverse instrument of a postcolonial construction aiming to establish the primacy of a state or territory within an ensemble still subject to a Western logic. This extreme danger is acutely perceived in the majority of "postcolonial" novels. Each time, the supposed rigor of the cartographer's work is matched to an imaginary geography that falls within the inner world of the characters.

For this reason, it seems essential to me to reposition space—each and every space, and the African space more than any other—in a time perspective. Taking inspiration from the work of Gilles Deleuze and Felix Guattari on strata and from Henri Lefebvre on "layered"[56] spatiality, I pointed out in my essay on geocriticism the need for a "stratigraphic" study that takes into account space and time in their interactions. Sometimes the relationship is "natural": temporal sedimentation occurs freely in a given space. In Africa, the relationship is biased. The work of sedimentation was interrupted, or at least disturbed, by a rereading imposed from the outside. The spatial impact of temporal manipulation consisting of erasing the indigenous references in order to replace them with new benchmarks is considerable. One of the most immediate indicators of this instability appears in the toponymy. Regardless of where the colonizer settles down, he is always eager to change the names of places or to adapt the existing names to his own language, in an effort to change the territorial configuration of the sites he occupied. In Africa, the map has been redrawn with effects that are still felt today, at a time when several attempts to return to the original have been launched. In South Africa, there is a gradual disappearance of Afrikaner names and a partial substitution of English place names. Thus the Northern Transvaal became the Northern Province in 1995, before being renamed Limpopo in 2001. The capital of Limpopo today has two names: Potgietersrus, and since 2003, Mokopane. These two names circulate almost interchangeably. But on what is this reciprocity based? Is it innocent? You should know that Potgietersrus was named after Pieter Potgieter, a Boer settler, a *Vortrekker* (a pioneer), killed near the site. As for Mokopane, the name was inherited from a local tribe, which had resisted the advance of the *Vortrekkers*. A form of onomastic legitimacy has been restituted to the place: those who tried to oppose an invasion have eventually restored their toponymic identity at the expense of those who robbed them of their land. This expected, necessary transformation occasionally plagued minds. In an article reprinted in the newspaper *Liberation*, on July 24, 2008, André Brink, the famous writer who spent part of his life fighting apartheid, expressed a legitimate anger after his nephew had been murdered by burglars in the eyes of his wife and his daughter. He added, however, "They lived north of Pretoria—or Tshwane in the new way of talking in vogue since the regime decided that only the ANC had a history in this country. (Nobody in their right mind would want to perpetuate the insulting names, witnesses of narrow-mindedness and racism of the past, but the current madness of changing names has reached an historical myopia, if not paranoia, which becomes an insult to the spirit that has made this new provision possible)."[57] Brink expresses himself with an understandable emotion, because in this process all is not historical myopia. On the same subject, Charles Maitland, a character in *Names* (1982), one of Don DeLillo's first novels, expresses nostalgia without nuance, unlike Brink. Shouting the accusation of "linguistic arrogance" at the changing of the names of Persia (Iran) and Rhodesia (Zimbabwe), the very conservative Maitland denounces, "Overthrow, re-speak. What do they leave us with? Ethnic designations. Sets of initials. The work of bureaucrats, narrow minds. I find I take these changes quite personally. They're a rescinding of memory."[58] The discussion between Charles Maitland and Hilaal would have certainly been rough. In the eyes of two characters, ethnic names and the abrogation of memory do not involve the same referents. For Charles Maitland, the categories of Mercator and his successors have become "natural." But they are not for David Keller, another character in the novel who notes that in the traditional Western vocabulary, white people "establish empires" and "scatter," while those with dark skin "unfold." And he wonders, "Why don't we say the Macedonians came sweeping out of Europe? They did. Alexander in particular. But we don't say that. Or the Romans or the Crusaders."[59]

In Kamila Shamsie's work, Raheen reflects on the hiatus between the different representations of the perceptible. To preserve the multiple realities of Karachi, the emblem of all postcolonial cities, she begins by refuting Karim's orthoscopic inclinations. In the mapping that he begins, she responds with a solemn warning: "That map is what marks you as an ex-pat and not as a Karachiite. People here don't talk in street names. And you never did either. You know that U2 song, 'Where the Streets have no name'. That's Karachi's song. Or, at least, the title is."[60] The Irish band's song is reminiscent of the intercommunity problem developed in Kamila Shamsie's novel. The streets with no name that Bono and U2 sing about are those of Belfast where the absence of referents promotes peace between Catholics and Protestants. Geographies revisited by an imagination imbued with the ideal of peace will find the debut of this application in the spaces to be recreated in postcolonial areas. At university, in a comparative literature exercise, Raheen parodies Italo Calvino's *Invisible Cities*. She conceives Zytrow city, where the streets have no names because everything is oriented according to the traders' stalls or actions as memorable as a visitor having jumped up and away. Zytrow is an exception:

> But if you leave Zytrow and forget its magic, you'll start listening to the poison of those who say all streets must have names. You'll join in the task of making directions easy for foreign travellers. And one by one, as you ink in your map, they disappear: the fruit seller, the ghosts, the friends you never said goodbye to. When the map is nearly done the cartographers will gather to celebrate. They'll say there's only one street remaining that needs a name. As they write the name and complete the map, someone tells you: before this, the inhabitants of Zytrow referred to it as the street where the boy leapt an incredible leap.[61]

In *Maps*, the situation is even more complex, as in the representation of spaces the human soul is solicited independently from reason. And Hilaal asks the fatal question to Askar: "Do you carve out of your soul the invented truth of the maps you draw? Or does the daily truth match, for you, the reality you draw and the maps others draw?"[62] The question takes Askar aback, but somehow the answer is part of the question: the truth of maps is an invented truth. Askar ends up by responding, "'Sometimes,' I began to say, 'I identify *a* truth in the maps which I draw. When I identify *this* truth, I label it as such, pickle it as though I were to share it with you . . . I hope, as dreamers do, that the dreamt dream will match the dreamt reality—that is, the invented truth of one's imagination. My maps invent nothing. They copy a given reality, they map out the roads a dreamer has walked, they identify a notional truth.'"[63] The soul is expressed here through the intervention of a dream, which traces an alternative geography. Its truth governs one of the possible worlds hosted in the postcolonial space. For both Somali characters, this world nevertheless incorporates an Ogaden free from Ethiopian influence.

The production of maps that are different from those imposed by Western civilization is sometimes facilitated by another relationship to writing. By "writing," I do not mean narration but conventional graphic signs. Michael Ondaatje writes about the Sinhala alphabet in *Running in the Family*: "Sanskrit was governed by verticals, but its sharp grid features were not possible in Ceylon. Here the Ola leaves which people wrote on were too brittle. A straight line would cut apart the leaf and so a curling alphabet was derived from its Indian cousin."[64] It is not difficult to imagine that the drawing of a map in Sri Lanka would have been incompatible with the vertical (and horizontal) nature of the borders. Sinhala cartography should accommodate the fragility of Ola leaves, which once endured short Buddhist writings. The maps are strange, foreign, producing strangeness. Coming from a tradition that is Occidental, although the first of them were Middle Eastern, they were called to fix into place the world of men, while all the while placing it at a distance. The European mapped the territory of Somali, Sri Lanka, and Pakistan. But these latter, in turn, adopted a cartographic point of view, tinged with a new form of objectivity—an objectivity that had become relative, like the Euclidean system upon which the mappemondes had been accustomed to rely. And in this oscillatory context, the strangeness continues to populate this lower world. Through the intervention of postcolonial literature, a less monolithic perception of place is possible, as clearly shown by Graham Huggan, because it causes "acceptance of diversity reflected in the interpretation of the map, not as a means of spatial containment or systematic organization, but as a medium of spatial perception which allows for the reformulation of links both within and between cultures."[65] The lesson of Deleuze and Guattari was adopted and assimilated. It would seem that the postmodern, postcolonial era has made a treasure out of the arrogant excesses of modernity that the great discoveries, or rather, "the major geographic inventions," opened as they closed space.

Cartography, Literature, Writing of Place

The map is now experiencing the favors of literature. Everyone remembers Robert Louis Stevenson and his *Treasure Island* (1883), or J. R. R. Tolkien and his sketch of Middle Earth. The map of the imaginary Yoknapatawpha County reigns at the end of *Absalom, Absalom!* (1936). But there is not just William Faulkner and the cartographers of fictitious countries. There is also Aritha Van Herk. In *No Fixed Address: An Amorous Journey* (1986), Arachne Manteia, the heroine of the novel, chooses to make love with Thomas on a pile of maps, which inspires her with some thoughts on how to orient her own biography. These maps "lead you into the past so easily, lead you through history to another frame of time. With these maps around, she would be able to transcend her own past, its rude, uneven measure, its gaps and horrors."[66] Maps maintain a privileged relationship with time. To be sure about that, we can refer to *Mason & Dixon* (1997) by Thomas Pynchon, or to *The Mapmaker's Opera* (2005) by Bea González, a Canadian novelist of Galician origin. The map has equally inspired painters. Vermeer and his globes, which decompartmentalize bourgeois homes, come to mind. Closer to home, there is *Map* (1961), in which Jasper Johns transmits his vision of the map of the United States. It is surprising to cite the series *Mappa* (1971–94), which includes the world maps of Alighiero Boetti, some of which have been woven by Afghan artisans. However, we will not return to the creations of Kathy Prendergast or to Guillermo Kuitca's series of cartographic acrylics whose title is *Untitled*. Painted on mattresses, these suggest motionless and incidental travel, sometimes erotic, dreamlike. The only possible routes are those traced out by the mattresses' buttons.[67]

There is no doubt that the map, the mapper, and mapping are fashionable. They have all three entered the field of postmodern metaphor. Everything is "maps," and "mapping" continues like crazy. If Mario Vargas Llosa became the winner of the Nobel Prize for Literature in October 2010, it is, according to the Swedish Academy "for his *cartography* of the structures of power."[68] As for Michel Houellebecq, he was awarded the Prix Goncourt a month later for *La Carte et le Territoire*, which portrays an artist working from Michelin maps. Yes, the map, the mapper, and mapping have entered the field of literature. Postcolonial output is heavily charged with maps: I mentioned a few Canadian novels (Van Herk, González), and I will cite an Australian novel (Murnane). Graham Huggan, who analyzed the cartographic discourse in contemporary Canadian and Australian literature, understands what is at play in the process: "the ironic and/or parodic treatment of maps as metaphors in post-colonial literary texts, the role played by these maps in the geographical and conceptual de/reterritorialization of post-colonial cultures, and the relevance of this process to the wider issue of cultural decolonization."[69] The map revisited is the deterritorialization agent of factious "certainties" conveyed by conventional Western reading. Provided that there are degrees in the postcolonial, it is recalled that some parts of the world have suffered the scourge of colonialism longer than others. Their desire to reconsider the configuration of place that was imposed on them is all the more vivid.

But the relationship between maps and text and images, in any form of fiction, is not new. In the relative void surrounding the generic questioning of maps,[70] we would gladly consider positing the exploration of boundaries where literature, maps, and atlases meet. In 1572, the first *Atlas* was placed at the intersection of disciplines. Mercator's positioning was not accidental. Given the interdisciplinary dynamic governing the cartographic undertaking, can we reasonably estimate that, from a certain point of view, the cartographer's work illustrates a genre with affinities to literary genericness? This link, although admittedly quite a stretch at the time of the GPS—has it ever existed? And if it did exist, why would it be stretched? Questioning the possible generic interactions between maps and literature assumes a dual approach. This would include pointing out what literary elements the cartographic text possessed (the delicate problem of *literaturnost*), but also determining the connection between the text and image—the latter having constituted the essence of the map since the Middle Ages. There is no need to activate here the complex relationship of the textual to the visual, which reflects the heterogeneous practices whose overlapping is often at random. In rhetoric, several types of couplings between text and image are established. Ekphrasis, the study of which is coming back in style, consists of rendering through the text an *objet d'art* that maintains a close relationship with it. This figure does not directly concern the map, which includes rather text in the image.[71] Indeed, it *no longer* concerns it, because once upon a time some maps were indeed textual. In Denys d'Alexandrie's *Périégèse*, the map consists of a sequence of names with a rectilinear layout that is nonplanar. This map refuses to be spread over the entire surface of the parchment or the wax; it is reduced to a line that attempts to restore order to the chaos of the world. Several maps of Late Antiquity and the Middle Ages that proposed routes have invested the crease between line and plane. In the fourth century, the famous *Tabula Peutingeriana*, a scroll 6¾ meters long and 34 centimeters wide, described the Roman Empire in all its depths, from Iberia to India, focusing on place names and on the network of roads with *Urbs* at the heart. The line was always present, but it is materialized in an image that remained ancillary.

The Hereford mappemonde confirms this view. It says, "*Omnia legenda quam pingenda.*"[72] This moto means, a priori, to "reproduce the readable rather than paint it." Commenting on this strange maxim, José Rabasa notes that "any pictorial representation requires an explanatory legend. So, instead of offering a way to determine the direction of the land, the written text should deepen the mirage and project desire."[73] It is as if the real and desire were all thrown together, the real as desire. On the maps of Matthew Paris, an English Benedictine monk of the thirteenth century, it was truly the image that had taken over. But the illuminations, summarily glossed, referred to the main cities leading from London to Dover and to Apulia, and from Apulia to the Holy Land; the cities were arranged in columns according to a specific device proper to the text. Between the fourth and thirteenth centuries, the hierarchy was reversed. The image had imposed its primacy to the text, at least for maps. Do not forget that maps were still read aloud at the time of St. Ambrose, a contemporary to the artisan of the *Tabula Peutingeriana*. People were not satisfied by just *looking* at maps.

Given this trend, the figure of ekphrasis seems less appropriate than the allegory to indicate the connection between the visual area and the textual line, because, in the allegory, according to José Rabasa, "literary space is transposed in visual representation."[74] For centuries, maps were subject to this mechanism of visualization. According to another definition, that of Bernard Dupriez, allegory is a "literary image whose —phore [comparing] is applied to the theme [the comparison], not globally as in metaphor or the figurative comparison, but piece by piece, or at least with personification."[75] The comparison here is a territory, an extract of the vast geographical world being entered. The comparison itself can vary. The situations are, however, not all equivalent. In some cases, the mapping allegory is classic and goes all the way to personification. We all have in mind *Europa Regina*, Bucius Johannes's map of Europe (1537), popularized by Sebastian Münster (1580), in which the continent adopts the rigid shape of a crowned woman whose head represents Spain, and the bottom of her dress Russia. Asia, for its part, took on the form of Pegasus, the winged horse, in Heinrich Bunting's 1581 representation. Other examples exist, such as *Leo Belgicus* by Aitzinger Michael (1583), a peaceful lion representing the Southern Netherlands after the split with the northern provinces of Spain. From the late nineteenth century, allegorized maps, close to caricatures, represented Scotland, England, Garibaldian Italy, Denmark,[76] and so on, in view of readings of a political nature. The allegory is sometimes the cryptic imaging of a text.

Cartography transmutes into text and image a reading of the world that is never unequivocal. The cartographer is not an anonymous and neutral body, but an author, a craftsman, sometimes an artist who models a dreamed vision of the human environment, employing an alchemist's approach—covering unknown territories with gilt that thereby become palatable. He delivers a work, just like a writer. Apparently, he sets himself the goal of fitting the entire world in his product. He aims to *transpose* it as is, according to a natural referent, while the writer, like any artist developing an imaginative representation, *transfigures* it. But in truth, the cartographer triggers a global dynamic where artifice and the natural feed each other, intertwine, and are even sometimes confused. While claiming to reproduce a model—the real "objective" that the geographical vision subsumes under it—it creates a territory. It is the entire world that enters into the subtle allegory supplied by the cartographer. The map is susceptible of being read like a book. And the atlas is a book. In it, the distance between the referent and its representation is supposed to be zero or negligible. It is an illusion that no cartographer legitimizes. The multiple variations of projection systems are there to prove it. Even photographic maps establish a plurivocal "reality." Objectivity is a myth that only ideological discourse feeds. And we know the real is the Other of discourse.[77] The map does not establish an absolute language; it standardizes difference in a language that is specific to each of its authors. As Michel de Certeau says, "refusing the fiction of a meta-language that unifies everything is to reveal the relationship between *limited* scientific procedures and that which they *lack* of the 'real' in what they address. It is to avoid the illusion, which is necessarily dogmatic and inherent to the discourse that pretends to believe it is 'adequate' to the real."[78] Or, indeed, cartography does not conform to reality, even if only because the "real" is not brought about within a single model or an imposed, stabilized hierarchy.

As the story told by the fictional narrator, the map offers a credible vision of the world. It creates a *possible world*. The mental map drawn, in readers' minds, by Don Quixote of his wanderings across the English Channel is a subjective view of Spain that ignores the *réalème*.[79] This is permissible, we will say, as it does not hold the literature, or Cervantes, or even less so Don Quixote responsible for drawing a consistent landscape. However, the *doxa* do not legitimize the margin of hesitation in maps that are presumed to reflect reality. Or, at best, they are the depleted substitute of that which they synthesize. Imagination is everywhere, and everything is simultaneously real and imaginary. The cartographic line will not help. At most, it testifies to a desire for order that is frustrated by the variability of the world and its representations. The infinity of space again evades the desire of enclosure that embodies place. In *Histoire de l'infamie, histoire de l'éternité* (1935), Jorge Luis Borges founded a College of Cartographers who decided to come up with a map of the empire that had not only the shape but also the size of the empire. Reducing the entire world to a representation is, however, excluded for the simple reason that mapping is a language and as such, it is imperfect. The map of the College of Geographers eventually disintegrates. Soon, we find nothing more than scattered remains. Luckily enough. For the map of the empire, as understood by Clement Rosset in *Le Démon de la tautologie* (1997), is a perfect example of tautology. It ensures that "all we can say about one thing is eventually reduced to simple enunciation, or re-enunciation, of the same thing."[80] Here the map aims to enclose the world on an interpretation of the space that should make a self-referential, autotelic, tautological commonplace.

The map maintains some affinity with the vast register of literary genres, the novel in particular. According to Michel de Certeau, it charges the name it postures with predicates, while the story gets the proper name according to a logic that assumes "a maximum shortening of the route and the distance between the functional foyers of the narration."[81] This compression results in a closing of loopholes. The same phenomenon is observable in cartography, but the referentiality of the name is more problematic. Like Greek tragedies, the maps have had a hard time traversing the centuries, materially. It is hard for place names, too, which follow political and colonial fluctuations. There exists a close relationship between mapping and postcolonial studies, which makes the map a popular document in literature. The cartographic denomination is attracting the attention of colonial theorists more and more. The prime meridian passes through Greenwich, and the Mercator Atlas forged a model whose layout has never been doubted, except after the fall of the great colonial empires. Maps have long been driven by a dynamic that put names in the center and text on the margin—a text that was used to fill the gaps in knowledge, which rendered up more of a fantasy than a reality yet to be built. The text is the legend, in short, in every sense of the term: what must be read, but also what we would like to believe. This marginal text deserves careful consideration because it establishes a contact zone between literature itself and a supposedly scientific discourse that poses the "objective" reality. The literariness of the map appears here in all its immediacy.

Place beyond Measure

The enclosure of space is the perfect culmination of the process of mastering/measuring that cartography applies to the immeasurable. This dynamic leads to an overall expansion of place and hence the extreme restriction and compression of the unknown. Ultimately, Seneca was wrong. Admittedly, Jason and the Argonauts had upset the balance of the world by venturing into a vibrant new space, but they only began a movement that would lead to transforming space into place and the world into a vast territory that is deeply striated. The erasure of the remaining blanks on the map reflects this appropriation. What Jason began in legendary times and Columbus continued in the early modern age was completed somewhere in the Congo, at the very end of the nineteenth century. Right in front of Marlow, the hero of *The Heart of Darkness* (1899), the last lingering part of the virgin map was colored in. Or the next-to-last, because Joseph Conrad hadn't seen everything. He was a few decades short of missing out on seeing this or that plot of land, a fragment of ice, or a mountaintop yield to the desire for filling in a part of humanity that was particularly turbulent: the one that is occupied by the Western world. All in all, the overall expansion of place corresponded to the irrepressible need to colonize the new in order to first reassure and then to boast. It should also comply with the unquenchable desire to take possession of the Other, to possess this Other in every sense of the polysemous predicate. This enormous deception has never ceased, without ever really emotionally or physically moving the modern West. And the West has never been other than modern and colonizing. In an essay titled *La Mobilisation infinie* (1989), Peter Sloterdijk summarizes the aporia of Western modernity carried by the blind force that is consubstantial to it. "There is," he writes, "a time called modernity, only because the power which impels Western men to act was able to make such a strong impression on itself that it found the courage to proclaim the organization of the world by its own action."[82] This action is recorded in "the movement toward more movement" in "a kinetic utopia."[83] From the point of view of the many who gave form and flesh to this aspiration, what was at stake was the unilateral stabilization of the world. It seems needless to say that the proclamation of this ultimate stasis should sanction the end of the process of territorialization of space, the consecration of an absolute location, the beginning of a *Pax Occidentalis* mode. This passage reflects the culmination of the modern as a secular Second Coming, triumphing "immeasurable powers" and, according to Sloterdijk, consecrating a "form which turns to the bearable, the imaginable, the elucidated, the circumvented."[84] Toward rationality, to sum it up in one word. But rationality involves the "mastered" relationship of the real to "measured" space and "the real is itself what is measurable, calculable, classifiable, thinkable."[85] By sheer chance, or due to a need that I would be hard pressed to explain, the Russian language expresses this idea with admirable clarity. *Mir* means both "peace" and "world" (or "universe"). *Mir* designates, according to Slavoj Žižek, author of *Welcome to the Desert of the Real* (2002), "the closed universe of the pre-modern peasant village community, evidently inherent in the idea that the entire universe is a harmonious whole in the image of the self-regulating peasant village."[86] Is modernity but the simple amplification of this primordial villager era? Between the world and the site of the village, is the difference reduced to a question of scale? Is the essence of One coextensive to that of the Other? The teleological projection of a global and harmonious world, or rather a harmonized world, feeds the memory erased from a germinal seed located in the steppe or elsewhere. The Russian language explicates what remains implicit somewhere else, that the village, the initial foyer, would be the *omphalos* and the world an extension progressing by concentric circular blows. It will surprise no one, therefore, that the modern individual, heir of this vision where the village and the world meet and overlap, can experience the feeling of suffocating in a space saturated with places, in a "global village," as if all paths of modernity led to the Gutenberg Galaxy! Sloterdijk expresses the overflowing dizziness that sanctions the incompatibility of space and the present moment: "In a world that has become to itself non-datable and non-narrative-able, every now is too narrow and wide for itself, the lack of space turns directly to the anguish of open spaces."[87] It is therefore necessary to empty out or find folds whereby to redeploy imagination.

But all the same, the path between *mir* and America has been heavily trod! So many efforts to reduce by all means the world space into a global village! So many civilizations and peoples stripped off the map to satisfy the *omphalos* complex plaguing the West! Could we just solve this complex? Everything suggests not. The world runs on the margins and beyond the vision that it inspires. It is cantilevered with that which it represents as rigid and thus with that which has any unique representation. The essence of space is inaccessible, because it transgresses the limits of the visible and the master-able. Place is nothing more than an example of space among others. Space is *im-mense*. That's why all the maps are different, why the representation of space is partial and, of course, inherently biased. In *La Communauté qui vient* (1990), Giorgio Agamben has noted with his usual sagacity that in German, the "example" is *Beispiel*. But the *Bei-Spiel* is "what plays next to," while in Greek, it was *paradeigma*, *paradeigma*, "which shows off next to." This playful and showy eccentricity is proper to place, and consequently the entire map is a collection of places destined to remain indefinitely open. And doubts about the stability of things and the world are permissible, "as the proper place of the example is always beside itself, in the empty space where its unspeakable and unforgettable life takes place."[88]

Let's take yet another brief detour to Australia. A few years before Bruce Chatwin would have headed to the bush, Gerald Murnane, an Australian writer who deserves to be better known, sent the character of an apprentice movie director to the deep country to shoot a film on the enigmatic existence of the inhabitants of the plains. *The Plains* (1982) is also the title of the novel. To better understand the specific context in which he is operating, the man is led to meditate on the nature of space. He quickly concludes that the projections of the real are a spiritual geography and that "obviously the plains do not coincide with any other known Australian land,"[89] though everyone, even the local residents, bind themselves to find landmarks "in the disturbing terrain of the mind."[90] The man is lost. He experiences the worst trouble conceiving the plot of the film. The discussions he overhears around him bring him no relief. The landowners usually meet in a tavern. Five of them share their views on the best way to understand space. The fifth, lamenting the end of the era of explorers, has spent years drawing the map of his land and naming places distinguished uniquely by himself. "Then in his later years he locked up all his notes and all his maps and invited all those who wanted to do so to explore the same site as him and make descriptions. When the various descriptions were compared, their discrepancies revealed the different qualities of each individual."[91] The third owner adds his two cents to the conversation. According to him, "the mission of the explorer is to postulate the existence of a country beyond the known world." And the ideal explorer is therefore an artist, who could account for this looming beyond the known. If this work were produced, the table companion would prefer to inquire into a new country: "I will go in search of places that are beyond the painted backgrounds, places that artists know that they are barely capable of indicating."[92] One can easily imagine the growing dismay of the director, for whom sectors of the real are collapsing and the proliferation of perspectives are thrown off balance. He solves the situation in his own way. He points a camera without any film in it to take the most mundane shots.

When addressing the question of the real—and of the reality of places—as a "treaty of idiocy" (idiocy initially being related to a singularity), Clement Rosset seems to be leaning in the direction of Giorgio Agamben in referring to the exemplary, the inherently sampled nature of things and places, the state of places. He could have even started to comfort Murnane's distraught director. After quoting Lucretius and Dante complaining, one about the uncertainty of time and place, and the other about the hassles that inspire the right track prone to error, Rosset is interested in the character of Geoffrey Firmin in *Under the Volcano* (1947) by Malcolm Lowry. Firmin, a former dismissed consul, tried unsuccessfully to escape the influence of alcohol in a small town in Mexico. The story is well known, and it is unnecessary to return to it. What catches the attention of Rosset is the limp of the diplomat, who accompanies Yvonne, as "somehow, anyhow,"[93] they moved on. The hobbling along of the alcoholic Geoffrey Firmin accompanied by Yvonne becomes symptomatic of what could be the entry into space of the whole of humanity, even if it is sober. Humanity too, facing the straight lines that maps idealize, hobbles along from one point to another, in a permanent *Bei-Spiel*. As Rosset said, "it is not 'in any way' (anyhow) which leads to a 'certain way' (somehow), that is to say, precisely to something that is not anything at all, in any way, but on the contrary to this here reality and no other, it has to be this way and not any other way. Total uncertainty and sheer determination are never confused with each other."[94] This reality characterizes the place in space. Certainly, it is not the reality that a globalizing, determining standard isolates. It is rather a reality that accompanies the myriad of possible variations of space and the multitude of escapades that the straight line provokes (even though it should make of space a global place). Place cannot be anything other than a metaphor for space, such as the example, the metaphor is that which brings together without identifying. Space is never really *there*; it's beyond, below, beside. *There* is just a place, which never coincides with space. There is an empty space, and it cannot be qualified. As somehow anyhow, it is in permanent transgression, in *transgressiveness*. For Rosset, it could be an *extravagance*, which, according to him, affects the Antigone of Sophocles as well as Lowry's consul. The extravagance makes of one and the other, and virtually any individual, a *pantoporos* being "capable of taking all roads, including the prohibited paths."[95]

In doing so, he braves the monotony of reality and the limitations of place. He opens up a path in space and reopens a space that nothing could hinder anyway. The maps here are powerless or "bombastic." They stage the devouring of a reality that, according to Rosset, "ceases to be an external reference and thus becomes both internalized and absent: so completely absent from the real world, outside of speech,"[96] somewhere between "interference and empty-handedness."[97] The *pantoporos* attitude consecrates the approximation and the approach relative to the detriment of the permanent and the absolute position. It is perhaps the only *plausible* one, which would tear the modern or postmodern individual away from the feeling of enclosure brought about by *localized* space, the space that maps, lines, and ridges put into place.

Notes

Foreword

1. See Bertrand Westphal, *Geocriticism: Real and Fictional Spaces*, trans. Robert T. Tally Jr. (New York: Palgrave Macmillan, 2011), especially 111–47.
2. See Bertrand Westphal, "Îles dalmates: L'odysée des îles," in *L'Œil de la Méditerranée* (La Tour d'Aigues, France: Éditions de l'Aube, 2005), 177–98.
3. Westphal, *Geocriticism*, 126–27.
4. Eric Prieto, "Geocriticism, Geopoetics, Geophilosophy, and Beyond," in Robert T. Tally Jr., ed., *Geocritical Explorations: Space, Place, and Mapping in Literary and Cultural Studies* (New York: Palgrave Macmillan, 2011), 22.
5. See Michel Foucault, *The History of Sexuality, Volume I: An Introduction*, trans. Robert Hurley (New York: Random House, 1978), 57–73.
6. Gilles Deleuze and Félix Guattari, *A Thousand Plateaus*, trans. Brian Massumi (Minneapolis: University of Minnesota Press, 1987), 500.
7. Here one might think of Marlowe's "blank space[s] of delightful mystery" in Joseph Conrad's *Heart of Darkness*. In "Geography and Some Explorers," Conrad praised the "honest" maps that left unknown or unexplored territories blank, rather than filling them with fanciful creatures or allegorical illustrations. Of course, what is left unsaid is the degree to which even scientific maps are projecting figural representations that stand in for "real" places, and these are no more or less real in an objective sense than the dragons or sea serpents of antique charts.
8. See, for example, Bertrand Westphal, *Spatiality* (London: Routledge, 2013), 79–86.
9. See Gianni Vattimo, "Dialectics, Difference, Weak Thought," in Gianni Vattimo and Pier Aldo Rovatti, eds., *Weak Thought*, trans. Peter Carravetta (Albany, NY: SUNY Press, 2012), 39–52.
10. Erich Auerbach, *Dante: Poet of the Secular World*, trans. Ralph Manheim (New York: New York Review of Books, 2007), 133.

Introduction Notes

1. [Translator's note: the word chosen by B. Westphal was *occire*, an archaic term for "slaying" that reinforces the role of the Occident by the repetition of *occ*.]
2. [Translator's note: The word chosen by B. Westphal is *côte*, which can mean "coast" or "rib." It is a wordplay for a gendered geographical metaphor.]

3. Zygmunt Bauma, *Identité*, trans. Myriam Denneby (Paris: L'Herne, 2010 [2004]), 101.
4. Bertrand Westphal, *Geocriticism: Real and Fictional Spaces*, trans. Robert Tally Jr. (New York: Palgrave, 2011).
5. Isabelle Autissier, "Postface," in *Ulysse et Magellan . . .* , Mauricio Obregón, trans. Marianne Saint-Amand (Paris: Autrement, 2003 [2001]), 117.
6. Michel Serres, *Esthétiques sur Carpaccio* (Paris: Le Livre de Poche, 2005 [1975]), 90.
7. [Translator's note: *métrise* is a Deleuzian play on words that combines the concepts of "measuring" and "mastering," here expressing the idea that there is a mastery of space through its measurement.]
8. Nicole Lapierre, *Pensons ailleurs* (Paris: Gallimard, 2006 [2004]), 195. This nice title, as indicated by Lapierre, is taken from Montaigne's *Essays* (Part III, Chapter 4).

Chapter 1

1. Auguste Bouché-Leclercq, *Histoire de la divination dans l'Antiquité* (Paris: Jérôme Million, 2003 [1879]), 600. The author adds, "This accommodation is not moreover, without prior example. The Romans also had their *Jupiter Lapis*, a flint that the Fetials carried with them, and a *Jupiter Terminus* which was riveted to the Capitol, like the *omphalos* of Delphi."
2. Varron, *De Lingua latina*, VII, 17, trans. D. Nisard (Paris: Dubochet, Le Chevalier et Cie, 1850), 528. The life of the very controversial Désiré Nisard, to whom we are indebted for this translation of book 7 (only book 6 has appeared in a recent edition of "Belles Lettres") deserves to be written about in a book. Éric Chevillard did so and the work is titled *Démolir Nisard* (Paris: Minuit, 2006).
3. Plutarch, *Sur la disparition des oracles*, 409^E, in *Dialogues Pythiques*, trans. Robert Flacelière (Paris: Belles Lettres, 1974), 100. In English: http://thriceholy.net/Texts/Oracles.html.
4. Ibid., 409^F
5. Lyrics by Ismael Serrano, "Km.0," song extracted from the album *Los Paraísos desertios*, TRAK, Madrid, 2000.
6. According to Geoffroy de Monmouth, a twelfth-century author of a work titled *Historia Regum Britanniae*, Brutus was the grandson of Aeneas. After much discussion, Brutus and his troops settled down in the island of Albion, to which the Trojan leader gave his name, *Bretagne*. He then built the city named *Troia Nova*, "The New Troy," the site of which was going to be London. This legend is one of many that follow in the long series of urban cosmogonies inspired by Greek mythology after the Trojan War.
7. Isidore De Séville, *Étymologies* (Paris: Belles lettres, 1981), 14: 3, 21.
8. Jahweh said to the Prophet Ezekiel (5:5), "This *is* Jerusalem: I have set it in the midst of the nations and countries *that are* round about her." *The Holy Bible Containing the Old and New Testaments: King James Version* (Cambridge: Cambridge University Press, 1995).

9. Samuel Y. Edgerton Jr., "From Mental Matrix to Mappamundi to Christian Empire: The Heritage of Ptolemaic Cartography in the Renaissance," in *Arts and Cartography: Six Historical Essays*, ed. David Woodward (Chicago: University of Chicago Press, 1987), 26.
10. Aristotle, *Physique*, trans. Pierre Pellegrin (Paris: Garnier-Flammarion, 2000), 202.
11. Ibid., 202–3. Available in English at http://classics.mit.edu/Aristotle/physics.4.iv.html, trans. R. P. Hardie and R. K. Gaye.
12. Giorgio Agamben, *Stanze. Parole et fantasme dans la culture occidentale*, trans. Yves Hersant (Paris: Payot & Rivages, 1998 [1977, 1992]), 13.
13. Ibid.
14. Ibid., 14.
15. The substantive *croisade*, which evolved from the ancient provincial *crozata* and the Spanish *cruzada*, comes later. It is not used until the fifteenth century. Certain synonyms can, however, be recognized, such as *croiserie*, *croisée*, or *croisement*.
16. See Agamben, *Stanze*, 219. The *stanza*, in Italian, is a "living room." An apartment has any number of *stanze*.
17. The word originally derived from Phoenician (*baitu-lilah*, "the residence of a God") and has passed through the Greek *baitulos*.
18. Predrag Matvejević, *Bréviaire méditerranéen*, trans. Evaine Le Calvé-Ivicevic (Paris: Fayard, 1992 [1987]), 139.
19. In his book (ibid.,141), Matvejević mentions a map that he found in Monastir, in Tunisia, on a camel's skin. In the center, it was not oriented toward Mecca but directly to the Ka'ba. The map was the work of Al-Sharfi, who worked at Sfax in the sixteenth century. Here, the *omphalos* takes on the appearance of a square that pivots on its axis and presides in the center of a disc in which the world is reduced to a thin circumferential band. The Ka'ba is linked to the circumference by 32 rays that seem to accentuate the emptiness of the world and that hardly interrupt a few scattered holy sites.
20. See François de Medeiros, *L'Occident et l'Afrique (XIIIe-XVe siècle)* (Paris: Karthala, 1985), 57–59.
21. O. H. Prior, ed., *L'Image du monde de Maître Gossouin* (Paris: Libraririe Payot, 1913), 103.
22. Prior, in his introduction to *Image du monde*, comments: "Aaron is probably the city named Aren on the map of Pierre Alphonse, a Jew from Huesca, who wrote around 1110. The form 'Arim' is found in a manuscript of *Image du Monde*, and makes this a likely assumption. This city, Miller said, in the middle of the earth, on the edge of the habitable world, is, according to the Arab legend, the refuge of demons and the throne of Iblis. This place, also called Aryn or Arym, is already mentioned by the Arabs in the ninth century. On a twelfth-century Persian map, it is shown as being in the middle of the earth. In the West, we often find that name in the thirteenth century" (38).
23. Pierre Alphonse, however, does not pass for a model of virtue or tolerance. His hypothesis appears in the *Dialogus contra Judaeos*, in which he attacked the Jewish and Muslim religions, so as to consolidate in a polemic outbidding his recent conversion to the Christian faith.

24. Daniele Del Giudice, *Horizon mobile*, trans. Jean-Paul Manganaro (Paris: Seuil, 2010 [2009]), 115.
25. Ibid., 115–16.
26. Ibid., 117.
27. Kikkawa Koretaru, cited in Augustin Berque, *Ecoumène. Introduction à l'étude des milieux humains* (Paris: Belin, 2000), 34.
28. Berque, *Ecoumène*, 39.
29. Ibid., 60.
30. See Ibid., 171. In French, we could read an article by Iinuma Jirô titled "La Logique spatiale dans l'agriculture japonaise" in *L'Espace géographique* 9 (1981): 143–48. The main idea summarized by the author has been theorized since the sixth century in a treaty on Chinese agronomy written by Jia Sixie: the *Qi Min Yao Shu*, which, among other things, inspired Charles Darwin.
31. Ricardo Padrón, *The Spacious World: Cartography, Literature, and Empire in Early Modern Spain* (Chicago: University of Chicago Press, 2004).
32. Ute Schneider, *Die Macht der Karten. Eine Geschichte der Kartographie vom Mittelalter bis heute* (Darmstadt: Primus Verlag, 2006), 27.
33. Paul Zumthor, *La Mesure du monde: Représentation de l'espace au Moyen Âge* (Paris: Seuil, 1993), 325.
34. Égerie, *Journal de voyage (Itinéraire)*, I, 19, 5, trans. Pierre Maraval (Paris: Cerf, 1982), 205.
35. Ibid., I, 2, 1–2, 123, 125.
36. D. K. Smith, *The Cartographic Imagination in Early Modern England: Re-Writing the World in Marlowe, Spenser, Raleigh and Marvell* (Aldershot: Ashgate, 2008), 34.
37. Alain Corbin, *Le Territoire du vide. L'Occident et le désir du rivage 1750–1840* (Paris: Flammarion, 1990 [1988]), 62.
38. Ibid.
39. Mary B. Campbell, *The Witness and the Other World: Exotic European Travel Writing. 400–1600* (Ithaca, NY: Cornell University Press, 1988), 24.
40. Smith, *The Cartographic Imagination*, 3.
41. Ibid., 83.
42. Erich Auerbach, *Mimésis. La représentation de la réalité dans la littérature occidentale,* trans. Cornélius Heim (Paris: Gallimard, 1977 [1946, 1968]), 125.
43. Ibid., 129.
44. Smith, *The Cartographic Imagination*, 36.
45. Benjamin de Tudèle, *Voyages de Rabbi Benjamin, fils de Jonas de Tudèle, en Europe, en Asie et en Afrique, depuis l'Espagne jusqu'à la Chine: où l'on trouve plusieurs choses remarquables concernant l'histoire et la géographie et particulièrement l'état des Juifs au douzième siècle,* trans. Jean-Philippe Baratier (Amsterdam: n.p., 1734), 156.
46. The four invariants of Hillah's description are such as in the *Rihla* of Ibn Jubayr. In one as in the other, Hillah had well-stocked markets, palm-tree enclosures, a pontoon bridge, and a large population. See, for example, the Spanish translation of the text by Ibn Jubayr: *A través del Oriente (Rihla)*, trans. Felipe Maíllo Salgado (Madrid: Alianza Literaria, 2007), onwards from page 334.

47. Ibn Battūta, *Voyages*, trans. C. Defremery and B. R. Sanguinetti (Paris: La Découverte/Poche, 1997), 1: 431.
48. In 1298, Marco Polo was captured by the Genoese, who competed at sea with the Republic of Venice, during the Battle of Curzola. He remained in prison until the summer of 1299.
49. Marco Polo, *Le Devisement du monde*, ed. A.-C. Moule and Paul Pelliot (Paris: La Découverte, 1998), 2:422.
50. Ibid., 423.
51. See Battūta, *Voyages*, 3: 255–67.
52. See more on this subject in the article by Ananda Abeydeera: "Taprobane, Ceylan ou Sumatra? Une confusion féconde," *Archipel* 47 (1994): 87–124.
53. *Viaggio di Nicolò di Conti veneziano, scritto per messer Poggio florentino*, in Giovanni Battista Ramusio, *Navigazioni e Viaggi*, ed. Marica Milanesi (Turin: Einaudi, 1978–88), 804. Digital version: *Progetto Manuzio*, *www.liberliber.it*.
54. *Andrea Corsali fiorentino allo illustrissimo principe e signor il signor duca Lorenzo de' Medici, della navigazione del mar Rosso e sino Persico sino a Cochin, città nella India, scritta alli XVIII di settembre MDXVII*, in Giovanni Battista Ramusio, *Navigazioni e Viaggi*, 474.
55. Ezekiel 28:13, in *The Holy Bible Containing the Old and New Testaments King James Version*.
56. Italo Calvino, *Les Villes invisibles*, trans. Jean Thibaudeau (Paris: Seuil, 1996 [1972]), 55.
57. Ibid., 69.
58. Ibid., 189.
59. Ibid., 31.
60. Ignace de Loyola, *Exercices Spirituels. Texte définitif (1548)*, trans. Jean-Claude Guy (Paris: Seuil, 1982), 69.
61. Gaston Bachelard, *La Poétique de l'espace* (Paris: Presses Universitaires de France, 1989 [1957]), 27.
62. Ignace de Loyola, *Exercices Spirituels*, 75.
63. Voir Benito Pelegrīn, *Figurations de l'infini. L'âge baroque européen* (Paris: Seuil, 2000), 78.
64. Christian Jacob, *La Description de la terre habitée (Périégèse) de Denys d'Alexandrie* (Paris: Albin Michel, 1990), 26. About Denys d'Alexandria, C. Jacob added, "Looking at a world map or reading the description, is to travel while standing still . . . Beating all the obstacles, you can travel in a straight line or roam the earth in endless arabesques" (ibid., 27).
65. [Translator's note: *lieux* means "place."]
66. According to the useful indications provided to me by my colleague Romain Garnier, *Rum* comes from the Indo-European root *ru*, which means "to open." In Latin, the root was perpetuated by *rus, ruris*, "the countryside" (from whence comes "rural," etc., in French). The German term *Raum* has maintained the element of "opening." The adjective *geräumig* also means "spacious."
67. It is found in Italian. The historian Dino Compagni, a contemporary of Dante and a Florentine like him, used it in a temporal sense in his *Cronica*, written

between 1310 and 1312, while in the *Volgar Crescenzi*, an agricultural treatise written around 1350, it referred to spatial meaning.
68. In German, *spazieren*, or *spazieren gehen*, means concretely "to go for a walk."
69. Paul Zumthor, *La Mesure du monde*, 51.
70. Nathaniel Harris, *Mapping the World: Maps and Their History* (San Diego: Thunder Bay Press, 2002), 43.
71. Smith, *The Cartographic Imagination*, 25.
72. Elizabeth Hill Boone, *Stories in Red and Black. Pictorial Histories of the Aztecs and Mixtecs* (Austin: University of Texas Press, 2000), 126.
73. Ibid., 164.

Chapter 2

1. Paul Zumthor, *La Mesure du monde: Représentation de l'espace au Moyen Âge* (Paris: Seuil, 1993), 52.
2. Vladimir Jankélévitch, *L'Aventure, l'ennui, le sérieux* (Paris: Aubier, 1963), 26. Ulysses is "an adventurer by force and a home-body by vocation."
3. Paul Zumthor, *La Mesure du monde*, 33.
4. Ibid., 61.
5. In 2003, 80 percent of it was destroyed during a violent earthquake that shook the Kerman province and especially the city of Bam.
6. Jacques Le Goff in *Popoli*, 2, 837–38, cited in Zumthor, *La Mesure du monde*, 240.
7. Zumthor, *La Mesure du monde*, 61–62.
8. [Translator's note: Here Westphal refers to Gilles Deleuze's wordplay of *métriser* and *maitriser*.]
9. Augustin Berque, *Écoumène, Introduction à l'étude des milieux humains* (Paris: Belin, 2000), 23. The author borrows this phrase from Jean-François Pradeau, "Être quelque part, occuper une place. *Topos* et *chôra* dans le *Timée*," in *Les Études philosophiques* 3 (1995): 396.
10. Berque, *Écoumène*, 24.
11. Ibid., 25.
12. Jean Echenoz, *Le Méridien de Greenwich* (Paris: Minuit, 1979), 248.
13. See David Harvey, *The Condition of Postmodernity* (London: Blackwell, 1990). In his essay, Harvey developed the concept of time-space compression.
14. Antonio Prete, *Trattato della lontananza* (Torino: Bollati Boringhieri, 2008), 40.
15. Revelation 12:18–13:1, *The Holy Bible Containing the Old and New Testaments King James Version* (Cambridge: Cambridge University Press, 1995).
16. Alain Corbin, *Le Territoire du vide. L'Occident et le désir du rivage 1750–1840* (Paris: Flammarion, 1990 [1988]), 25.
17. Ibid., 12.
18. Aeschylus, *Les Perses*, trans. Danielle Sonnier and Boris Donné (Paris: Flammarion, 2000), 749–50.143.
19. Dante, *La Divine Comédie. Enfer*, XXI, 117, trans. Jacqueline Risset (Paris: Flammarion, 1992 [1985]), 243.

20. We can add to this beginning of a list the Spaniard Fernando S. Llobera. In *El noveno circulo* (2005), he staged a serial criminal who was inspired by the configuration of Dante's Hell. With *Dante Alighieri ei delitti della Medusa* (2000), the Italian Giulio Leoni began a cycle with three subsequent volumes—namely, *The Third Heaven Conspiracy* (*I delitti del mosaico*, 2004), *The Conspiracy of Mirrors* (*I delitti della luce*, 2005), and *The Crusade of Darkness* (*The crociata delle tenebre*, 2007)—that have been translated into French by Belfond. Dante here embodies the character of a detective, as Aristotle in a series of novels by Canadian comparatist Margaret Doody.
21. See Ildefonso Cerdá, *Teoría General de la Urbanización* (Madrid: Imprenta Espagñola, 1867).
    The transformation of Barcelona under the leadership of Cerdá, considered one of the fathers of modern urbanism both for his practical achievements and for his theoretical work, has been the object of a remarkably romantic description in *La Ville des prodiges* (1986) by Eduardo Mendoza.
22. See Virgil, *L'Énéide*, V, 755, trans. Maurice Rat (Paris: Garnier-Flammarion,1965), note 1260, 336: "Some Latin grammarians even have it come from the word *urbs*, from the old word *urbum* or *urvum*, handlebar of the plow . . . Pomponius wrote about it, see Digest, L, XVI, 239: '*Urbs ab urvo appellata is urvare is aratro definire*'. This etymology is anything but certain."
23. Ibid., V, 755, 124–25. I owe this reference to Augustin Berque, *Écoumène*, chapter 8: "La cité."
24. For more information, see the analyses of Massimo Cacciari in *La città* (Villa Verrucchio, Italy: Pazzini, 2004), 14 and onwards from page 14.
25. Jorge Luis Borges, "Le Dernier Voyage d'Ulysse," in *Neuf essais sur Dante*, trans. Françoise Rosset (Paris: Gallimard, 1999 [1982]), t. 2: 842.
26. There are not many studies devoted to this book. In the abstract to her article "*Places Far from Ellesmere*: Aritha Van Herk's Tesselated Territory," published in *Études canadiennes* 25, no.47 (1999): 177–90, Claire Omhovere comments on the term: "Aritha van Herk sees in *Places Far From Ellesmere* a *geografictione*. The *portmanteau* word indicates that this autobiography tends to steer clear of traditional approaches. To execute the work, she chose to hide the 'I' in order to take interest in the context that gives it meaning. As a result, she focuses on the forms of writing that have helped transform the places of origin in a promising colonial territory . . . Van Herk bends the representation of the North to make a feminine territory out of it, when it was originally cut to fit the measures of the fantasies of its male explorers."
27. Aritha Van Herk, *Places Far from Ellesmere* (Red Deer, Canada: Red Deer College Press, 1995 [1990]), 130.
28. Ibid., 130–31.
29. [Translator's note: *orizonte* is Corsican for "horizon."]
30. See Michel Collot, *L'Horizon fabuleux. XIXe siècle* (Paris: José Corti, 1988), 31–32.
31. See Pierre de Ronsard, Marc-Antoine Muret, *Les Amours, leurs Commentaires*, eds. Christine de Buzon and Pierre Martin (Paris: Didier Érudition, 1999 [1553]), 91–92, for the commentary by Muret, of which here is the beginning: "From the

horizon, In whatever place we have discovered, it seems that we see like a circle, which on all sides arrests and achieves our view. Such circles are called in Greek Horizons."
32. Attilio Momigliano, *Commento a* La Divina Commedia, "Purgatorio" (Firenze: Sansoni, 1982 [1945]), 269: "In *Purgatory* astronomic poetry began to abound."
33. Among them, we can read Danièle James-Raoul and Claude Alexandre Thomasset, eds. *Dans l'eau, sous l'eau: le monde aquatique au Moyen Âge* (Paris: Presses de l'Université de Paris-Sorbonne), 2002.
34. In reality, the "Atlantic" was introduced a little later in the English language. It was in 1601 that an "Atlantick Ocean" appeared in the translation that Philemon Holland did of Pliny's *Histoire naturelle*. We would have to wait until 1744 for the "Atlantic Ocean" to be drawn on a map by Emmanuel Bowen, in the sense that we now recognize it.
35. Æthiopia was much larger than the current Ethiopia. It was the "Country of Black Men" and corresponded in substance to everything we knew or thought we knew about sub-Saharan Africa. See more about this topic in François de Medeiros, *L'Occident et l'Afrique*. The author makes a careful examination of the various occurrences of the term throughout his essay.
36. Patrick Deville, *Pura vida. Vie & mort de William Walker* (Paris: Seuil, 2004), 17.
37. "Nicaragua," *Diccionario enciclopedico hispano-americano* (Barcelona: Montaner y Simon, 1898).
38. Deville, *Pura vida*, 36.
39. Prete, *Trattato della lontananza*, 46.
40. Deville, *Pura vida*, 39.
41. Prete, *Trattato della lontananza*, 47.
42. Predrag Matvejević, *Bréviaire méditerranéen*, 125.
43. Herodotus, *L'Enquête*, IV, 42, trans. Andrée Barguet (Paris: Gallimard, 1990 [1964]), t. 1, 376.
44. Manlio Brusatin, *Histoire de la ligne*, trans. Anne Guglielmetti (Paris: Flammarion, 2003 [1993, 2001]), 65.
45. Giovanni Battista Ramusio, *Navigazioni e Viaggi*, ed. Marica Milanesi (Turin: Einaudi, 1978–88), 1456.
46. Could this be the Vinland where, after Erik the Red and his Vikings have stayed, the Bishop of Greenland was mentioned by Adam of Bremen? The conjecture was brought up early on, but it was immediately challenged. Washington Irving discusses it, before dismissing it, in *A History of the Life and Voyages of Christopher Columbus* (1828).
47. Ramusio, *Navigazioni e Viaggi*, 1457.
48. Ibid., 1458.
49. Michel Collot, "L'Horizon du paysage," in *Lire le paysage, lire les paysages* (Saint-Etienne: Publications du CIEREC, 1984), 122.
50. Ibid., 123.
51. Ibid., 126. [Translator's note: this sentence uses wordplay around the term *fable*, such as "*fabuleux*" (fabulous) and "*affabulation*" (fabrication), to emphasize the link between fable and the creation thereof.]
52. Giorgio Agamben, *Nudités*, trans. Martin Rueff (Paris: Payot & Rivages, 2009), 29.

53. Ibid., 28. In the original version, Agamben uses the term *tenebra* (in the singular) in preference over *tenebre* (in the plural). The singular is possible in a literary Italian.
54. Ibid., 25.
55. Pascal Quignard, *Boutès* (Paris: Galilée, 2008), 75.
56. Pindare, *Olympiques*, III, 44–45, trans. Aimé Puech (Paris: Belles Lettres, 1970), 56.
57. Michel de Certeau, *L'Écriture de l'histoire* (Paris: Gallimard, 2002 [1975]), 180.
58. Dante, *La Divine Comédie. Purgatoire*, I, 44–45, trans. Jacqueline Risset (Paris: Flammarion, 1992 [1988]), 19. In English: http://www.gutenberg.org/files/8795/8795-h/8795-h.htm#link1.
59. Prete, *Trattato della lontananza*, 42.
60. Dante, *Purgatoire*, I, 13, 17. In English: http://www.gutenberg.org/files/8795/8795-h/8795-h.htm#link1.
61. Ibid., II, 14–15, 25. In English: http://www.gutenberg.org/files/8795/8795-h/8795-h.htm#link1.
62. Ibid., I, 130–32, 23. In English: http://www.gutenberg.org/files/8795/8795-h/8795-h.htm#link1.
63. Ibid., I, 102–5, 21. In English: http://www.gutenberg.org/files/8795/8795-h/8795-h.htm#link1.
64. Ibid., I, 23–24, 17. In English: http://www.gutenberg.org/files/8795/8795-h/8795-h.htm#link1.
65. [Translator's note: here there is a wordplay with *sidéreal* and *sidérant*, which means "amazing." In translation we lose the repetition of the prefix *sidér*-.]
66. Dante, *Purgatoire*, II, 101, 29. In English: http://www.gutenberg.org/files/8795/8795-h/8795-h.htm#link1.
67. Ibid., II, 104–5, 29. In English: http://www.gutenberg.org/files/8795/8795-h/8795-h.htm#link1.
68. Ibid., I, 4, 17. In English: http://www.gutenberg.org/files/8795/8795-h/8795-h.htm#link1.
69. Daniele Del Giudice, *Horizon mobile*, trans. Jean-Paul Manganaro (Paris: Seuil, 2010 [2009]), 216.
70. Ibid., 26–217.
71. See Dante, *La Divine Comédie. Paradis*, trans. Jacqueline Risset (Paris: Flammarion, 1992 [1985]), 33.133.
72. Youri Lotman, *La Sémiosphère*, extracted from *L'Univers de l'esprit*, trans. Anka Ledenko (Limoges: PULIM, 1999 [1966]), 108.
73. Ibid., 111.
74. Alejandro Gándara, *La media distancia* (Madrid: Alfaguara, 2008 [1984]), 159.
75. Julien Gracq, *Le Rivage des Syrtes* (Paris: Corti, 2009 [1951]), 68–69.
    It does not matter here if the narrator invents a variety of reed that, for botanists, does not exist. It has the gift of sexualizing places. Rather than the "ilve" one would indeed expect that the "hard stemmed reed" approaches the "Ulva," an algae with a suggestive name.
76. Ibid., 47.

Chapter 3

1. [Translator's note: The Symplegades are also known as the Clashing Rocks or the Cyanean Rocks.]
2. Apollonios de Rhodes, *Argonautiques*, II, 417–18, trans. Emile Delage (Paris: Belles Lettres, 1974), t. 1, 197.
3. The story almost received the recognition it deserved. Homer refers to the ship Argo and to the Symplegades in Song XII, v. 69–72, of the *Odyssey*, when Circe shows Ulysses the way. In the *Fourth Pythian Ode*, Pindar has paid more attention to the trip of Jason and his men, but his story remains succinct, and, even though he mentions crossing the Symplegades, he does not reflect on the Argonauts' experience of discovering unknown space.
4. Apollonios de Rhodes, *Argonautiques*, II, 320, t. 1, 191.
5. Ibid., II, 333–34, t. 1, 192.
6. Ibid., II, 607–9, t. 1, 206.
7. Predrag Matvejević, *Bréviaire méditerranéen*, trans. Evaine Le Calvé-Ivicevic (Paris: Fayard, 1992 [1987]), 177–78.
8. Valerius Flaccus, *Argonautiques*, IV, 711–15, trans. Gauthier Liberman (Paris: Belles Lettres, 1997), 142–43.
9. Homer, *Odyssée*, X, 508, trans. Frédéric Mugler (Arles: Actes Sud, 1995), 185.
10. Ibid., XI, 15, 189.
11. Seneca, *Médée*, trans. Charles Guittard (Paris: GF Flammarion, 1997), 55.
12. Ibid., 56.
13. Ibid., 69.
14. Ibid., 56.
15. Dante, *Enfer*, IV, trans. Jacqueline Risset (Paris: Flammarion, 1992 [1985]), 141, 55. In English: http://www.gutenberg.org/files/8789/8789-h/8789-h.htm #link4.
16. Ibid., IV, 144, 55.
17. Matvejević, *Bréviaire méditerranéen*, 141.
18. Fernando Pessoa, *Message* [1934], in *Poèmes ésotériques. Message. Le Marin*, trans. Michel Chandeigne and Patrick Quillier (Paris: Christian Bourgois, 1988), 129.
19. Ibid.
20. Maria Bellonci, *Marco Polo* (Milano: Rizzoli,1989 [1982]), 10.
21. Ibid.
22. Ibid., 252.
23. Ibn Fadlan, cited by Stéphane Yerasimos, "Introduction," in Ibn Battūta, *Voyages*, trans. C. Defremery and B. R. Sanguinetti (Paris: La Découverte/Poche, 1997), 2:42.
24. Battūta, *Voyages*, 2:228–29.
25. Matvejević, *Bréviaire méditerranéen*, 156.
26. Ibid., 157.
27. On his tomb, which is located inside the Basilica of St. Peter, the following epitaph is engraved: "*Novi orbis suo aevo inventi Gloria*"—that is, "To him belongs the glory of the discovery of the New World."
28. Al-Umarī, *Recueil des sources arabes concernant l'Afrique occidentale du VIIIe au XVIe siècle*, trans. Joseph Cuoq (Paris: CNRS, 1975), 275–78.

29. Maurice Delafosse, *Haut-Sénégal-Niger* (Paris: Maisonneuve & Larose, 1972 [1912]).
30. [Translator's note: *griots* are members of a West African hereditary caste responsible for maintaining the oral history of the tribe.]
31. Camara Laye, *Le Maître de la parole* (Paris: Plon, 1978), 258.
32. The Afrocentric approach was not born in the seventies. It arose in the 1920s in the work of Leo Wiener, Norbert's father, who founded his hypothesis, among other assumptions, on the supposed linguistic analogies between and Amerindian and West African languages.
33. Columbus's error having fizzled, the common vocabulary leads us still today to talk about "Indians." Sometimes a scientific veneer is added to the error by referring to the "Amerindians." In the United States, the formula *Native Americans* is used, but the Native Americans, except for collective qualifications, use the names of the various nations to which they belong. For my part, I rely on the Quebecois use of terms that refer to *aboriginals* and *First Nations*.
34. [Translator's note: The French city of Limoges has hosted *Les Francophonies en Limousin*, a Francophone festival in which Francophone authors, playwrights, and musicians have gathered to present their works since 1984. See more at http://www.lesfrancophonies.com/index.html.]
35. Gaoussou Diawara, *Abubakari II* (Carnières: Lansman, 1992); Gaoussou Diawara, *Avec 2000 bateaux il partit . . . La Saga du Roi Mande Bori* (Oslo: Skyline, 2000); Jean-Yves Loude, *Le Roi d'Afrique et la reine mer* (Arles: Actes Sud, 1994); Alfred Bosch, *El Atlas furtivo*, trans. Pau Pérez (Barcelona: Planeta, 2007 [1998; 1999]).
36. Diawara, *Abubakari II*, 19.
37. Djibril Tamsin Niane, *Sounjata ou l'épopée mandingue* (Paris: Dakar, Présence Africaine, 1994 [1960]), 152.
38. Loude, *Le Roi d'Afrique et la reine mer*, 54.
39. Ibid., 157.
40. Bosch, *El Atlas furtivo*, 269.
41. Loude, *Le Roi d'Afrique et la reine mer*, 77.
42. Ibid., 80.
43. Diawara, *Avec 2000 bateaux il partit*, 68.
44. Ibid., 177.
45. Ibid., 222–23.
46. Ibid., 28.
47. Ibid., 34.
48. Cheikh Anta Diop, *Nations nègres et culture: de l'antiquité nègre égyptienne aux problèmes culturels de l'Afrique noire d'aujourd'hui* (Paris: Dakar Présence Africaine, 1999 [1954]).
49. For a brief, bare-facts presentation, see for example, Chiu Ling-Yeong, "Zheng He: Navigator, Discoverer and Diplomat," in *Wu Te Yao Memorial Lectures 2000* (Singapore: Unipress, 2001).
50. *Sanbao's Expedition to the West Sea*: This narrative was reissued in China (Pékin, Huaxia, 1995) but has never been translated into a European language.

51. Gavin Menzies, *1421, l'année où la Chine a découvert l'Amérique*, trans. Julie Sauvage (Paris: Intervalles, 2007 [2002]), 55.
52. See Mauricio Obregón, *Ulysse et Magellan* . . . , trans. Marianne Saint-Amand (Paris: Autrement, 2003 [2001]), 114–15.
53. Dante, *La Divine Comédie. Enfer*, XXVI, trans. Jacqueline Risset (Paris: Flammarion, 1992 [1985]), 108–9, 243. In English: http://www.gutenberg.org/files/8789/8789-h/8789-h.htm#link4.
54. François Rabelais, *Gargantua* (Genève: Droz, 1970), 69. In English: http://www.gutenberg.org/files/1200/1200-h/1200-h.htm.
55. François Rabelais, *Pantagruel* (Geneva: Droz, 1965), 171. In English: http://www.gutenberg.org/files/1200/1200-h/1200-h.htm.
56. Ibid.
57. Erich Auerbach, *Mimésis, La représentation de la réalité dans la littérature occidentale*, trans. Cornélius Heim (Paris: Gallimard, 1977 [1946, 1968]), 271.
58. Augustin Berque, *Ecoumène, Introduction à l'étude des milieux humains* (Paris: Belin, 2000), 70.
59. Edmundo O'Gorman, *L'Invention de l'Amérique. Recherche au sujet de la structure historique du Nouveau Monde et du sens du devenir*, trans. Francine Bertrand-Gonzalez (Québec: Presses Universitaire Laval, 2007 [1958]), 84.
60. Ibid., 96.
61. Ibid., 145.
62. The Acts of the Apostles 7:55–58, in *The Holy Bible Containing the Old and New Testaments King James Version* (Cambridge: Cambridge University Press, 1995).
63. Michel Serres, *Esthétiques sur Carpaccio* (Paris: Le Livre de Poche, 2005 [1975]), 87.
64. Frank Lestringant, *L'Atelier du cosmographe ou l'image du monde à la Renaissance* (Paris: Albin Michel, 1991), 47.
65. François Hartog, *Mémoire d'Ulysse. Récits sur la frontière en Grèce ancienne* (Paris: Gallimard, 1996), 96. "It is, it seems, with Eratosthenes in the 3rd century BC, that the term *geographos* was introduced, as one who draws or describes the land, the author of a treatise on geography or cartographer. Previously it was 'périégète,' author of a 'journey' or 'tour' in the inhabited world."
66. Paul Zumthor, *La Mesure du monde: Représentation de l'espace au Moyen Âge* (Paris: Seuil, 1993), 227.
67. Ibid., 232.
68. Ibid., 229.
69. See Hartog, *Mémoire d'Ulysse*, 46–49, where this question is closely studied, without being related to the myth of Io.

Chapter 4

1. Yves Berger, "Préface," in Alvar Nuñez Cabeza de Vaca, *Relation de voyage*, trans. Bernard Lesfargues and Jean-Marie Auzias (Arles: Actes Sud, 1979), 12.
2. Ibid., 184.
3. Ibid., 151.

4. Paul Zumthor, *La Mesure du monde: Représentation de l'espace au Moyen Âge* (Paris: Seuil, 1993), 162.
5. Gilles Deleuze and Félix Guattari, *Capitalisme et schizophrénie 2. Mille Plateaux* (Paris: Minuit, 1980), 471.
6. Ibid., 473–74.
7. Thomas Gomez, *L'Invention de l'Amérique. Mythes et réalités de la conquête* (Paris: Flammarion, 1992), 60.
8. Bertrand Westphal, *Geocriticism: Real and Fictional Spaces,* trans. Robert T. Tally Jr. (New York: Palgrave, 2011).
9. José Rabasa, *L'Invention de l'Amérique. Historiographie espagnole et formation de l'Eurocentrisme,* trans. Claire Forestier-Pergnier and Eliane Saint-André-Utudjian (Paris: L'Harmattan, 2002 [1993]). For the American side, see William Boelhower "Inventing America: The Culture of the Map," which appeared in the *Revue Francaise d'Etudes Americaines,* 13, no. 36 (April 1988): 211–24.
10. Rabasa, *L'Invention de l'Amérique,* 18.
11. Fernan Perez de Oliva's book, mentioned by José Rabasa, has long been forgotten. It was republished by the Instituto Caro y Cuervo in Bogotá in 1965, in an edition by Juan José Arrom. In 1993, following the festivities, it appeared at the Presses de l'Université de Cordoba, by Pedro Ruiz Pérez. For my part, I quote the version from the Biblioteca Virtual Miguel de Cervantes, www.cervantesvirtual.com.
12. Roland Barthes, *L'Aventure sémiologique* (Paris: Seuil, 1985), 125.
13. Alonso de Ercilla, *La Aracauna,* XV, 13, trans. Alexandre Nicolas (Paris: Utz, 1993), 281.
14. Rabasa, *L'Invention de l'Amérique,* 84.
15. Apollonios de Rhodes, *Argonautiques,* IV, 885–919, t. 3, 108–9.
16. Pascal Quignard, *Boutès* (Paris: Galilée, 2008), 27, 28.
17. Homer, *Odyssée,* X, 395–96, 181.
18. Giorgio Agamben, *L'Ouvert. De l'homme et de l'animal,* trans. Joël Gayraud (Paris: Payot & Rivages, 2002), 59.
19. Ibid., 60.
20. In fact, Agamben's "anthropological machine" falls in the wake of the "mythological machine" (*macchina mitologica*) of his compatriot Furio Jesi, Germanist and anthropologist. Opting for a political reading, Jesi deconstructed the myth based on the denunciation of a propaganda mechanism from a speech hiding its origins, naturalizing its historical significance, and supplying a genuine "mythological machine." See for example Furio Jesi, *La Fête et la machine mythologique,* trans. Fabien Vallos, with an introduction by Giorgio Agamben (Paris: Mix, 2008 [1977]).
21. Pierre d'Ailly, *Imago Mundi, Latin Text and French Translation of the Four Cosmographic Treaties of d'Ailly and Marginal Notes by Christopher Columbus* (Gembloux: Duculot, 1930), 240–41. Haly is the astrologer Abū l-Hasan 'Ali ibn Abi l-Rijāl, who lived between the end of the tenth century and the beginning of the eleventh century. He is more known in Europe under the name of Haly Abenragel.
22. [Translator's note: Here Westphal uses a wordplay, "dé-lire," which combines the verb *délirer* or "to become delirious" with *dé-lire* or "dis-read"/ "un-read."]

23. Fernando Pessoa, *Message*, [1934] in *Poèmes ésotériques. Message. Le Marin*, trans. Michel Chandeigne and Patrick Quillier (Paris: Christian Bourgois, 1988), 132.
24. Luis de Camões, *Les Lusiades*, V, 50, trans. Roger Bismut (Paris: Belles Lettres, 1980), 109.
25. Jean-Yves Loude, *Lisbonne. Dans la ville noire* (Arles: Actes Sud, 2003), 82.
26. Ibid.
27. *Périple* of Hannon, fragment 18.
28. Moreover, it sometimes occurs that the creatures are male and inhabiting a universe prey to sexual dysfunction. In a passage of his *Descriptions* (I, 23, 5–6), Pausanias reports the testimony of Eupheme of Caria. After the winds had ejected his ship in the outer sea, Eupheme landed beyond the limits of the known world, an island in the Satyrides archipelago. It was inhabited by savage men, red and fitted with a tail barely shorter than that of a horse. When they saw the ship, the Satyrs boarded the ship to capture the women, spreading panic. Faced with the threat, the sailors disembarked a woman (barbarian), who was delivered to the worst outrages.
29. See Gomez, *L'Invention de l'Amérique*, 120–21.
30. See Odile Gannier, *Les Derniers Indiens des Caraïbes. Image, mythe et réalité* (Matoury, Guyane: Ibis Rouge, 2003), 205–8.
31. Homer, *Odyssée*, X, 289, 293–301, 178. In English: http://classics.mit.edu/Homer/odyssey.10.x.html.
32. Ricardo Padrón, *The Spacious World: Cartography, Literature, and Empire in Early Modern Spain* (Chicago: University of Chicago Press, 2004),194.
33. Ibid.
34. Ibid.,195.
35. De Ercilla, *La Aracauna*, XV, 13, 281.
36. Ibid., X, 7, 206.
37. Michel de Certeau, *L'Écriture de l'histoire* (Paris: Gallimard, 2002 [1975]), 279.
38. In *L'Occident et l'Afrique (XIIIe-XVe siècle)* (Paris: Karthala, 1985), 30, François de Medeiros has inventoried the different names of the continent. For the Romans, *Africa* was initially a province including Carthage. Then *Africa* was juxtaposed to *Aethiopia*, the first being white Africa and the second black Africa. The boundaries between the two were blurred. *Africa* became Africa, a small portion of which was known, when the world was divided into three parts.
39. De Certeau, *L'Écriture de l'histoire*, "Avant-propos à la seconde édition," 9–10; Rabasa, *L'Invention de l'Amérique*, 41–55.
40. De Certeau, *L'Écriture de l'histoire*, 9.
41. Osvaldo Soriano, "Fútbol. Relatos épicos sobre un deporte que despierta pasiones," [1994] *Booket* 2317 (2010): 112.
42. Homer, *Odyssey*, XII, 186, 217.
43. Gonzalo Fernández de Oviedo, *Historia General y Natural de las Indias* (Madrid: Biblioteca de Autores Españoles, 1959), 17.
44. Ibid.
45. [Translator's note: Marianne is the female allegory of the French Republic, representing liberty, equality, and fraternity.]
46. Ibid., 20.

47. José Manuel Fajardo, *Lettre du bout du monde*, trans. Claude Bleton (Paris: Flammarion, 1998 [1996]).
48. See Francisco López de Gómara, *Historia General de Las Indias*, XIII (Caracas: Biblioteca Ayacucho, 1979), 28.
49. Fernando Iwasaki Cauti, *El descubrimiento de España* (Oviedo: Ediciones Nobel, 1996), 73.
50. Walter D. Mignolo, *The Idea of Latin America* (Malden, MA: Blackwell, 2005), 53.
51. Stefan Zweig, *Amerigo. Récit d'une erreur historique*, trans. Dominique Autrand (Paris: Le Livre de Poche, 1996 [1942]), 59.
52. Ibid., 121–22.
53. Albert Ronsin, *Le Nom de l'Amérique* (Strasbourg: La Nuée Bleue, 2006), 202–4.
54. See Louis-Jean Calvet, *Linguistique et colonialisme. Petit Traité de glottophagie* (Paris: Payot, 1974).
55. Julien Gracq, *La Forme d'une ville* (Paris: Corti, 1985), 2.
56. Xavier Moret, *América, América. Viaje por California y el Far West* (Barcelona: Península, 2001 [1998]), 221–22.
57. Francisco López de Gómara, *Historia General de Las Indias*, XIII (Caracas: Biblioteca Ayacucho, 1979), 299.
58. See Joyce Marcus, *Mesoamerican Writing Systems: Propaganda, Myth and History in Four Ancient Civilizations* (Princeton, NJ: Princeton University Press, 1992).
59. Serge Gruzinski, *Les Quatre parties du monde. Histoire d'une mondialisation* (Paris: Seuil, 2006 [2004]), 276–312. The author mentions other notable cases, such as Martín Ignacio de Loyola, the nephew of the founder of the Society of Jesus, or that of Salvador Correia de Sá, who worked between Brazil, Angola, and Mombasa.
60. Barron Field, cited in Paul Carter, *The Postcolonial Studies Reader*, eds. Bill Ashcroft, Gareth Griffiths, and Helen Tiffin (London: Routledge, 2001 [1995]), 404.
61. De Certeau, *L'Écriture de l'histoire*, 302.
62. Christopher Marlowe, *Tamerlan*, 2, V, 3, 159, trans. Philippe de Rothschild (Paris: Albin Michel, 1977), 243.
63. Ibid., 2, V, 3, 146–47, 243.
64. See D. K. Smith, *The Cartographic Imagination in Early Modern England: Re-Writing the World in Marlowe, Spenser, Raleigh and Marvell* (Aldershot: Ashgate, 2008),125–55.

Chapter 5

1. Bruce Chatwin, *The Songlines* (London: Jonathan Cape, 1987), 2.
2. Ibid., 13.
3. Ibid., 58.
4. Ibid., 56.
5. Vishvajit Pandhya, "Movement and Space: Andamanese Cartography," *American Ethnologist* 17, no. 4 (1990): 784.
6. Chatwin, *The Songlines,* 306.

7. Gilles Deleuze and Félix Guattari, *Mille Plateaux* (Paris: Minuit, 1980), 254.
8. Jules Laforgue, "Hamlet ou les suites de la piété familiale," in *Moralités légendaires, Œuvres completes* (Geneva: Slatkine, 1979), 13.
9. See Frank Lestringant, *L'Atelier du cosmographe ou l'image du monde à la Renaissance* (Paris: Albin Michel, 1991), 40.
10. Heiner Müller, *Paysage avec Argonautes*, in *Germania Mort à Berlin*, trans. Jean Jourdheuil and Heinz Schwarzinger (Paris: Minuit, 1985 [1982]), 18.
11. Carl Schmitt, *Le Nomos de la terre*, trans. Lilyane Deroche-Gursel (Paris: PUF, 2001 [1950–88]), 87.
12. Ibid., 89. Schmitt adds a little further, "Scarcely had the first maps and globes been crafted, scarcely had the early scientific representations of the true shape of our planet and a new world to the west been brought out, than the first global lines of partition and distribution were being drawn," ibid., 90.
13. Francisco López de Gómara, *Historia General de Las Indias*, XIII (Caracas: Biblioteca Ayacucho, 1979), 153.
14. Ibid.
15. Frances Gardiner Davenport, *European Treaties Bearing on the History of the United States and Its Dependencies, 1917–1937* (Washington, DC: Carnegie Institute, 1967), 220.
16. Hergé, *Les Aventures de Tintin. Le Trésor de Rackham le Rouge* (Brussels: Casterman, 2007 [1944]), 23.
17. Dana Sobel, *Longitude: The True Story of a Lone Genius Who Solved the Greatest Scientific Problem of His Time* (London: Fourth Estate, 1998 [1996]), 56.
18. René-Primavère Lesson, *Voyage autour du monde entrepris par ordre du gouvernement sur la corvette La Coquille*, cited in *Viajeros franceses a las Islas Canarias*, eds. Berta Pico and Dolores Corbella (La Laguna: Instituto de Estudios Canarios, 2000), XL.
19. Jean Echenoz, *Le Méridien de Greenwich* (Paris: Minuit, 1979), 10.
20. See Umberto Eco, *The Island of the Day Before* (1994). For his part, Jules Verne had long discussed the Date Line. In *Around the World in Eighty Days* (1873), the well-informed Phileas Fogg begins his world tour by the east in order to gain a day crossing the line.
21. Echenoz, *Le Méridien de Greenwich*, 10–11.
22. Ibid., 193.
23. See Osvaldo Soriana, *Fútbol*, "*Últimos días del arquero feliz. A un siglo del la invención del penal*," Relatos épicos sobre un deporte que despierta pasiones, [1994] *Booket* 2317 (2010): 35–39.
24. Blaise Pascal, *Pensées* (Paris: Gallimard, 1977), 87. In English: http://etext.lib.virginia.edu/etcbin/toccer-new2?id=PasThou.xml&images=images/modeng&data=/texts/english/modeng/parsed&tag=public&part=all.
25. Paul Zumthor, *La Mesure du monde: Représentation de l'espace au Moyen Âge* (Paris: Seuil, 1993), 313.
26. Brian Harley, *Maps in Tudor England* (London: British Library, 1993), 15. Cited in D. K. Smith, *The Cartographic Imagination in Early Modern England: Re-Writing the World in Marlowe, Spenser, Raleigh and Marvell* (Aldershot: Ashgate, 2008), 77.

27. Lestringant, *L'Atelier du cosmographe*, 183.
28. Walter D. Mignolo, *The Idea of Latin America* (Malden, MA: Blackwell, 2005), xiii.
29. Ibid.
30. Michel de Certeau, *L'Écriture de l'histoire* (Paris: Gallimard, 2002 [1975]), 365.
31. Elizabeth Hill Boone, *Stories in Red and Black: Pictorial Histories of the Aztecs and Mixtecs* (Austin: University of Texas Press, 2000), 165–66.
32. Brian Harley, *The New Nature of Maps*, ed. Paul Laxton (Baltimore: Johns Hopkins University Press, 2001), 144–45. Passing away prematurely in 1991, Harley is the proponent of a philosophy of mapping that relies on rhetoric and reading of Michel Foucault and Jacques Derrida, among others.
33. Ibid., 139.
34. Ibid., 188.
35. See ibid., 105.
36. Lestringant, *L'Atelier du cosmographe*, 165.
37. See Ute Schneider, *Die Macht der Karten. Eine Geschichte der Kartographie vom Mittelalter bis heute* (Darmstadt: Primus Verlag, 2006), 100.
38. See Manuel Francisco de Barros E Sousa, Vicomte de Santarém, *Essai sur l'histoire de la cosmographie et de la cartographie* (Paris: Maulde and Renou, 1849–52).
39. Peter Turchi, *Maps of the Imagination: The Writer as Cartographer* (San Antonio, TX: Trinity University Press, 2004), 37.
40. Harley, *The New Nature of Maps*, 99.
41. Michael Ondaatje, *Running in the Family* (Toronto: McClelland & Stewart Limited, 1982), 63.
42. Ibid.
43. Ibid.
44. Ibid., 61–62.
45. Harley, *The New Nature of Maps*, 56.
46. Law number 2005-158 dating from February 23, 2005 recognizes the nation and the national contribution in favor of repatriated French citizens. Cf. Art. 4, al. 2: "School programs are to recognize in particular *the positive role of the French presence overseas*."
47. Identifying recurrences of cartography in postcolonial literary production will be the subject of at least one book, which is still yet in part to be written.
48. Kamila Shamsie, *Kartographie* (Boston: Mariner Books, 2007 [2002]), 40.
49. Nuruddin Farah, *Maps* (New York: Penguin Books, 2000 [1986]), 229.
50. See Mark Monmonier, *Comment faire mentir les cartes. Du mauvais usage de la géographie*, trans. Denis-Armand Canal (Paris: Flammarion, 1994 [1991]).
51. Guillermo Kuitca, in Leah Ollman, "An Artist Finds His Place in the World," *Los Angeles Times*, June 11, 1995, 56. http://www.speronewestwater.com/cgi-bin/iowa/articles/record.html?record=5.
52. Shamsie, *Kartographie*, 24.
53. Whereas *Congo* is of precolonial origin (*Kongo*), *Zaïre* derives from a Portuguese word that, it is true, has its origins in *Nzere*, signifying "river" in the Kikongo language.

54. Tijan Sallah, "Banjul Afternoon," in *Dreams of Dusty Roads* (Colorado Springs, CO: Three Continents Press, 1993), 24. See Sylvie Coly, "La vision de l'Afrique dans la poésie sénégalaise et gambienne: Léopold Sédar Senghor, Lenrie Peters, Amadou Lamine Sall et Tijan M. Sallah" (PhD diss., University of Limoges, 2010), 91.
55. Farah, *Maps*, 270.
56. [Translator's note: The term used here is *feuilleté*, as in a *mille-feuille* pastry made of phyllo dough, which has many superimposed layers.]
57. André Brink, "Ténèbres à Midi," trans. Jean-Charles Burou, in *Libération*, July 24, 2008.
58. Don DeLillo, *The Names* (New York: Vintage, 1989 [1982]), 240.
59. Ibid., 260.
60. Shamsie, *Kartographie*, 133.
61. Ibid., 127.
62. Farah, *Maps*, 227.
63. Ibid., 228.
64. Ondaatje, *Running in the Family*, 83.
65. Graham Huggan, "Decolonizing the Map. Post-Colonialism, Post-Structuralism and the Cartographic Connection," in *Ariel*, 20, no. 4 (1989): 124.
66. Aritha Van Herk, *No Fixed Address: An Amorous Journey* (Toronto: McClelland & Stewart, 1986), 117.
67. See Giuliana Bruno, *Atlas of Emotion: Journeys in Art, Architecture and Film* (London: Verso, 2002).
68. *The Nobel Prize in Literature 2010—Prize Announcement*, http://nobelprize.org/nobel_prizes/ literature/laureates/2010/annoucement.html.
69. Huggan, "Decolonizing the Map," 123.
70. See José Rabasa, *L'Invention de l'Amérique. Historiographie espagnole et formation de l'Eurocentrisme*, trans. Claire Forestier-Pergnier and Eliane Saint-André-Utudjian (Paris: L'Harmattan, 2002 [1993]), 192: "As far as I know, there is no history of the atlas as a genre."
71. It may be noted, however, that in one of the oldest texts in the French language, *Roman de Thèbes*, written around 1150, there is a description of a world map painted on the silk wall of the tent of Adrastus, the prince of Argos, in the form of ekphrasis.
72. [Translator's note: "all things that must be read rather than painted."]
73. Rabasa, *L'Invention de l'Amérique*, 87.
74. Ibid., 43.
75. Bernard Dupriez, *Gradus. Les procédés littéraires* (Paris: U.G.E., 1984), 29.
76. See, for example, Nathaniel Harris, *Mapping the World: Maps and Their History* (San Diego: Thunder Bay Press, 2002),144–47.
77. See Michel de Certeau, *L'Écriture de l'histoire*, 38.
78. Ibid., 11.
79. [Translator's note: A *réalème* is a technical term in comparative literature, attributed to L. Dolezel, that refers to the relationship between represented space, its literalization, and its referent.]
80. Clément Rosset, *Le Démon de la tautologie* (Paris: Minuit, 1997), 48.

81. Michel de Certeau, *L'Écriture de l'histoire*, 133.
82. Peter Sloterdijk, *La Mobilisation infinie. Vers une critique de la cinétique politique*, trans. Hans Hildenbrand (Paris: Seuil, 2003 [1989]), 22.
83. Ibid., 56, 23.
84. Ibid., 261.
85. Ibid., 239.
86. Slavoj Žižek, *Bienvenue dans le désert du réel*, trans. François Théron (Paris: Flammarion, 2007 [2002]), 125.
87. Ibid., 297.
88. Giorgio Agamben, *La Communauté qui vient. Théorie de la singularité quelconque*, trans. Marilène Raiola (Paris: Seuil, 1990), 16–17.
89. Gerald Murnane, *The Plains* (New York: George Braziller, 1985 [1982]), 37.
90. Ibid., 40.
91. Ibid., 54.
92. Ibid., 56.
93. See Malcolm Lowry, *Under the Volcano* (London: Jonathan Cape, 1967 [1947]). The consul speaks with gravity about the activities of his half-brother to Yvonne "while, anyhow, they were on their way."
94. Clément Rosset, *Le Réel. Traité de l'idiotie* (Paris: Minuit, 1977), 13.
95. Ibid., 23.
96. Ibid., 105.
97. Ibid., 124.

Index

Abeydeera, Ananda, 169
Adam of Bremen, 55, 61, 172
Adorno, Theodore, 2
Aeschylus, 45, 51, 100
Agamben, Giorgio, 7, 17–19, 64–65, 71, 112–13, 120, 161–62, 167, 172–73, 177, 183
Aguirre, Lope de, 58, 103, 117
Alcaraz, Diego de, 106
Alexander III (Pope), 19
Alexander IV (Pope), 139
Allende, Salvador, 22
Almeyda, Alvaro, 124
Alphonse, Pierre, 21, 22, 167
Alphonse V, 231
Ambrose (Saint), 51, 157
Amerike, Richard, 129
Andahazi, Federico, 94, 167
Angelopoulos, Theos, 27
Aristotle, 18, 78, 110, 123, 167, 171
Arrien, 33
Auerbach, Erich, xvi, 29, 94, 165, 168, 176
Augustus, 14
Autissier, Isabelle, 4, 6, 166
Azores, 78, 91, 141
Azpilcueta, Martin de, 126

Bachelard, Gaston, 37, 169
Bacon, Roger, 21
Bakari II, Abu, xvii, 61, 82, 84–86, 88–90, 95, 128, 139, 175
Baratier, Jean-Philippe, 32, 168
Barros e Sousa, Manuel Francisco de, 148, 181
Barthes, Roland, 110, 177

Bath, Adelard de, 21
Battūta, Ibn, 30–32, 34, 36, 81–82, 85, 169, 174
Bauman, Zygmunt, 4
Beatus of Liébana, 15–16
Becerra, Diego de, 117
Becket, Thomas, 19
Bellonci, Maria, 80–81, 174
Benjamin, Walter, 48
Benoit, Claude, xvii
Berger, Yves, 105, 176
Bernières, Louis de, 27
Berosus the Chaldean, 123
Berque, Augustin, 23–24, 47, 54, 95, 168, 170–71, 176
Béthencourt, Jean de, 2
Bevis of Hampton (Beuves de Hantone), 28–29, 39
Blaeu, Joan, 121
Blanche of Castile, 17
Bodenhamer, David J., xi
Boelhower, William, 177
Boetti, Alighiero, 155
Bolívar, Simón, 59
Boone, Elizabeth Hill, 41, 146, 170, 181
Borges, Jorge Luis, 54, 159, 171
Bosch, Alfred, 86, 88, 175
Bouché-Leclercq, August, 11, 166
Bove, Giacomo, 21
Bowen, Emmanuel, 172
Brendan (Saint), 52–53, 61
Brink, André, 153, 182
Bruno, Giuliana, 182
Brusatin, Manlio, 61, 172
Buckle, Henry Thomas, 151
Bunting, Heinrich, 158

Buzon, Christine de, xvii, 171
Buzzati, Dino, 46

Cabeza de Vaca, Alvar Núñez, xiv, 103–9, 136, 176
Cabot, John, 62, 129
Cabral, Pedro Alvares, 127
Cacciari, Massimo, 171
Cain, Byron, 142–43
Calvet, Louis-Jean, 130, 179
Calvino, Italo, 36, 153, 169
Camões, Luis de, 92, 115–16, 178
Campanella, Tommaso, 34
Campbell, Mary B., 28, 168
Cardinal of Salzbourg, 36
Carpaccio, 4, 97–99, 166, 176
Carroll, Lewis, 3
Castillo, Alonso del, 104, 106
Cato of Utica, 67
Cato the Elder, 110
Caucasia, 80
Cauti, Fernando Iwasaki, 179
Cavendish, Thomas, 22
Cerdá, Ildefonso, 53, 171
Certeau, Michel de, 67, 120, 133, 145, 158–59, 173, 178
Cervantes, Miguel de, 158, 177
Charles I, 132
Charles II, 130
Charles V, 93
Chatwin, Bruce, 41, 135–36, 162, 179
Chrysoloras, Manuel, 34
Cicero, Marcus Tullius, 56, 66, 110
Clark, William, 148
Cleombrotus the Lacedaemonian, 13
Coelho, Duarte, 87
Coincy, Gautier de, 45
Coleridge, Samuel T., 133
Collot, Michel, 56, 63–64, 171–72
Columbus, Christopher, xiv, 2, 23, 31, 56, 60, 62, 78–79, 80–85, 87, 93, 95–99, 101–3, 105–6, 108–9, 113–15, 117, 124–25, 127–30, 132, 137, 160, 172, 175, 177
Coly, Sylvie, 182
Compagni, Dino, 169

Condé, Babou, 84
Conrad, Joseph, 160, 165
Corbin, Alain, 27, 50–52, 168, 170
Corneille, Pierre, 50
Corrigan, John, xi
Corsali, Andrea, 31, 169
Cortés, Hernán, 103–4, 117, 127, 130, 146
Cosa, Juan de la, 82, 114
Courbet, Gustave, 3
Covarrubias Horozco, Sebastián de, 25
Cranach le Jeune, Lucas, 12
Cresques, Abraham, 83, 86, 91
Cresques, Jafudá, 88
Croix, Jean de la, 38

d'Ailly, Pierre, 113–14, 177
d'Alexandrie, Denys, 38, 157, 169
Damiani, Pier, 38
Dante Alighieri, xvi, xvii, 4, 6, 13, 52–57, 66–71, 78, 86, 92, 96, 99, 116, 162, 165, 169–71, 173–74, 176
d'Anville, M. Jean-Baptiste Bourguignon, 148
Darwin, Charles, 168
d'Astypalée, Onésicrite, 32
Davenport, Frances Gardiner, 180
d'Avila, Theresa, 38
Dear, Michael, xi
de'Conti, Nicolò, 91
Delafosse, Maurice, 84, 175
Deleuze, Gilles, xiii, 5, 48, 107–8, 113, 136–37
DeLillo, Don, 153, 182
Demetrius, 13
Denon, Vivant, 28
Derrida, Jacques, xv, 181
Descartes, René, 39
de Tischbein, 12
Deville, Patrick, 58, 172
Diabaté, Bala, 89
Diagne, Pathé, 89
Dias, Bartolomeu, 115
Diawara, Gaoussou, 86–87, 89–90, 175
Diego, Juan, 103

Digby, Sir Kenelm, 141
Diodore of Sicily, 33
Diop, Cheick Anta, 90, 175
Doležel, Lubomir, 3, 182
Doody, Margaret, 171
Dorantes, Andres, 104, 106
d'Oria, Tedisio, 56
Du Bellay, Joachim, 99
Dupriez, Bernard, 157, 182

Earp, Wyatt, 40
Echenoz, Jean, 49, 142, 170, 180
Echevarría, Nicolás, 103–4, 106
Eco, Umberto, 19, 142, 180
Edgerton, Samuel Y., 16–17, 167
Egeria, 26, 28, 30
Eldred, John, 35
Elias, Amy, xi
Ely, Ron, 40
Empedocles, 110
Eratosthenes, 21, 99, 176
Ercilla y Zúñiga, Alonso de, 119
Erik the Red, 54
Etienne (Saint), 97–99
Etzlaub, Erhard, 22

Fadlan, Ibn, 81, 174
Fajardo, José Manuel, 124–25, 179
Farah, Nuruddin, xvii, 149–51, 181–82
Fari, Aljauma. *See* Ferrer, Jaime
Faulkner, William, 155
Fernandez, Peregrino, 121–22
Ferrariis, Theophilus de, 123
Ferrer, Jaime, 83, 88
Ferrer, Jaume. *See* Ferrer, Jaime
Field, Barron, 133, 179
Flaccus, Valerius, 75, 174
Flagg, James Montgomery, 121
Flores, Juan José, 59
Foigny, Gabriel de, 95
Foucault, Michel, xiii, 165, 181
François I, 130
Freising, Otto von (Bishop), 19
Freud, Sigmund, 119
Frontera, Palos de la, 78

Galileo Galilei, 141
Gall, James, 150
Gama, Vasco de, 36, 92, 114–15, 118
Gambia, Bakari Sidibe, 88
Gamboa, Pedro Sarmiento de, 22
Gándara, Alejandro, 69, 173
Gandhi, Mahatma, 15
Gannier, Odile, 178
Gaozhi (Hongxi), Zhu, 91
Garnier, Romain, 169
Gast, John, 121
Gatto, Alfonso, 17
Gerlache, Adrien de, 21
Gervase, Provost of Tilbury, 17
Ginzburg, Natalia, 17
Giudice, Daniele Del, 21–22, 68, 168, 173
Godard, Jean-Luc, 17
Godefroy of Bouillon, 16
Goethe, Johann Wolfgang von, 28
Gómara, Francisco López de, 125, 131, 140, 179–80
Gomez, Thomas, 109, 117, 177–78
González, Bea, 155–56
Gracq, Julien, 46, 70, 131, 173, 179
Gregoretti, Ugo, 17
Gruzinski, Serge, 131, 179
Guattari, Félix, xiii, 5, 48, 107–8, 113, 136, 138, 152, 155, 165, 177, 180
Guernes of Pont-Sainte-Maxence, 18

Habermas, Jürgen, 4
Haeckel, Ernst, 112
Haley, Alex, 151
Haly Abenragel, 113, 177
Hannon, 60, 116, 178
Harley, Brian, 144, 146–49, 180–81
Harris, Nathaniel, 40, 170, 182
Harris, Trevor M., xi
Harrison, John, 141
Hartog, François, 176
Harvey, David, 170
He, Ma. *See* He, Zheng
He, Zheng, 90–92, 95, 175
Hergé (Georges Prosper Remi), 180
Herodotus, 45, 51, 54, 60, 116, 120

Herzog, Werner, 58, 103
Heyerdahl, Thor, 84, 92
Hierro, 78, 141
Higden, Ranulph, 57
Holland, Philemon, 172
Homer, 12–13, 23, 54–55, 74, 76, 112–14, 118, 174, 177–78
Horozco, Sebastián de Covarrubias, 25
Houellebecq, Michel, 156
Huggan, Graham, 155–56, 182
Hugo, Victor, xv, 93

Iambulos, 33
Iborra, Juan Luis, 14
Idrissi, Muhammad Al-, 20, 92
Innocent IV (Pope), 33
Innocent VIII (Pope), 83
Irving, Washington, 172
Iwasaki, Fernando, 94, 125, 179

Jacob, Christian, 38, 169
Jankélévitch, Vladimir, 170
Jesi, Furio, 177
Jirô, Iinuma, 24, 168
Jodorowsky, Alejandro, 104
Johannes, Bucius, 157
John (Saint), 15
John II (King), 115
John of Cornwall, 57
Johns, Jasper, 155
Joinville, Jean de, 19
Jourdain of Sévérac, 34
Joyce, Jame, xii
Jubayr, Ibn, 32, 168
Juzay, Ibn, 82

Kaysersberg, Johann Geiler von, 30
Kazantzakis, Nikos, 27
Keita, Sundiata, 84–85
Ketchum, Jim, xi
Khaldun, Ibn, 83–84
Khan, Kublai, 36, 96, 114, 129
Kiney, Abbot, 131
Kinski, Klaus, 58
Klee, Paul, 48
Koretaru, Kikkawa, 23, 168

Kuitca, Guillermo, 150–51, 155, 181
Kupperman, Karen Ordahl, 147

Laforgue, Jules, 137, 180
Lambert, Johannes Heinrich, 150
Lapierre, Nicole, 7, 166
Las Casas, Bartolomé de, 82, 101
Laurel, Stan, 121
Laye, Camara, 84, 175
Lefebvre, Henri, 152
Le Goff, Jacques, 46–47, 102, 170
Leoni, Guilio, 171
Lery, Jean de, 145
Lesson, René-Primavère, 142, 180
Lestringant, Frank, 98, 137, 144, 148, 176, 180–81
Ling-Yeong, Chiu, 175
Llobera, Fernando S., 171
Llosa, Mario Vargas, 156
Lombardie, Ascelin de, 33
Longjumeau, André de, 33
Lotman, Yuri, 69, 173
Loude, Jean-Yves, 86–90, 115, 175, 178
Lowry, Malcolm, 162–63, 183
Loyola, Ignace de, 37, 169
Lucretius Carus, Titus,162
Luria, Sarah, xi
Lyotard, Jean-François, 2

Madeira, 78
Magellan, Ferdinand, 36, 91, 119, 127, 166, 176
Mājid, Ahmed ibn, 36
Malocello, Lancilotto, 56
Mandeville, Jean de, 19, 30–31, 34–35, 45, 57
Manheim, Ralph, xvi, 165
Manuel I Comnenus of Byzantium, 19
Maodeng, Luo, 91–92
Marcou, Jules, 128
Marcus, Joyce, 131, 179
Marignolli, Giovanni, 34
Marlowe, Christopher, 133–34, 165, 168, 179–80
Marsto, William M., 117
Martin, Pierre, 171

Matvejević, Predrag, 20, 60, 75, 78, 82, 167, 172, 174
Mauro, Fra, 23
Maximilien of Transylvania, 36
Medeiros, François de, 167, 172, 178
Mela, Pomponius, 33, 171
Menzies, Gavin, 91, 176
Mercator, Gerardus, 58, 80, 90, 114, 144, 150, 153, 156, 159
Metz, Gossuin de, 21
Michael, Aitzinger, 151
Mignolo, Walter D., 126–27, 144, 179, 181
Mitchell, Peta, xi
Moll, Hermann, 147
Momigliano, Attilio, 172
Monmonier, Mark, 181
Monmouth, Geoffroy de, 166
Montaigne, Michel de, 166
Montalvo, Garci Rodríguez de, 117
Montesquieu, Charles-Louis de Secondat, Baron de La Brède et de, 94
Moraru, Christian, xi
More, Thomas, 144
Moret, Xavier, 131, 179
Müller, Heiner, 137, 180
Münster, Sebastian, 57, 157
Muret, Marc-Antoine, 56, 171
Murnane, Gerald, 156, 162, 183
Musa, Kankou (Messe Melly), 83–85
Musa, Mansa, 85, 88

Narváez, Panfilo de, 104–5, 107
Nero, 78
Niane, Djibril Tamsin, 87, 175

Obregon, Mauricio, 91–92, 166, 176
O'Gorman, Edmundo, 96–97, 108–9, 176
Ojeda, Alonso de, 131
Oliva, Fernan Perez de, 109, 177
Olmec, 84
Omhovere, Claire, 171
Ondaatje, Michael, xvii, 148–49, 154, 181–82
Orellana, Francisco de, 117

Ortelius, 58
Ovid, 81, 100, 146
Oviedo, Gonzalo Fernández de, 93, 123–25, 131, 178–79

Padrón, Ricardo, 25, 119, 168, 178
Pandhya, Vishvajit, 136, 179
Paris, Matthew, 157
Pascal, Blaise, 143, 180
Pasolini, Pier Paolo, 17–18
Pausanias, 11–13, 178
Pavel, Thomas, 3
Pelegrīn, Benito, 169
Pencrych, Richard, 57
Perez, Domingo, 124
Pessoa, Fernando, 79, 174, 178
Peters, Arno, 150–51
Petrarch, 34
Pharaoh Necho, 60
Pilate, Pontius, 107
Pinart, Alphonse, 129
Pindar, 10–11, 66, 173–74
Pinochet, Augusto, 22
Pinzòn, Martin Alonso, 79, 128
Pizarro González, Francisco, 103, 127
Pizzigano, Zuane, 91
Plan Carpin, Jean de, 33
Plato, 18, 47, 125, 139
Plato of Tivoli, 21
Pliny the Elder, 31–32, 61, 125, 172
Plutarch, 13, 166
Pocahontas, 118, 132
Polo, Marco, 19, 23, 33–34, 36, 45, 80–81, 96, 130, 169, 174
Pompilius, Numa, 26
Portugal, Laurent de, 33
Potgieter, Pieter, 152
Poulet, Georges, 137
Poussin, Nicolas, 146
Pradeau, Jean-François, 47, 170
Prendergast, Kathy, 174, 151, 155
Prester John, 19, 30, 34, 45, 86, 148
Prete, Antonio, 50, 59, 67, 170
Prieto, Eric, xi, xii, 165
Prior, O. H., 167
Proconnèse, Aristéas de, 38

Proust, Marcel, 127
Ptolemy, Claudius, 20–21, 34–36, 78, 113, 145
Pujante, Domingo, xvii
Pynchon, Thomas, 155
Pytheas of Massalia, 60

Quignard, Pascal, 65, 111, 173, 177
Quint, Charles, 117

Rabasa, José, 109–10, 120, 157, 177–78, 182
Rabelais, François, 93–94, 176
Rada, Martín de, 132
Rada, Rodrigo de (Archbishop), 123
Raleigh, Sir Walter, 117, 134, 168, 179–80
Ramusio, Giovanni Battista, 61–62, 116, 169, 172
Ratisbonne, Petahiah de, 32
Rhodes, Apollonios de, 74–75, 111, 174, 177
Richard of Haldingham, 40
Richardson, Doug, xi
Richelieu (Cardinal), 140–41
Roger II (King), 20
Ronsard, Pierre de, 56, 171
Ronsin, Albert, 128, 179
Rossellini, Roberto, 17
Rosset, Clement, 7, 159, 162–62, 182–83
Rubens, Peter Paul, 12
Rubroek, Guillaume de, 33
Rückert, Friedrich, 22
Rustichello of Pisa, 33, 82

Saint-Bris, Thomas Lambert de, 128
Saint-Saëns, Camille, 12
Salinas, Martín de, 117
Sallah, Tijan, 151, 182
Salvatores, Gabriele, 27
Sanbao, Ma. *See* He, Zheng
Savage, Thomas S., 116
Schedel, Hartmann, 146
Schmitt, Carl, 138–40, 142, 180
Schneider, Ute, 148, 168, 181

Sefardi, Moshé (Pierre Alphonse, Pedro Alfonso), 21
Seko, Mobutu Sese, 151
Seneca, 77–78, 80, 86, 160, 174
Serrano, Ismael, 14, 166
Serrano, Yolanda García, 14
Serres, Michel, 4, 98, 166, 176
Séville, Isadore de, 32
Shamsie, Kamila, xvii, 149–50, 153, 181–82
Shaw, Flora, 151
Siciliano, Enzo, 17
Sinclair, Henry (Earl of Orkney), 62
Sloterdijk, Peter, 7, 160–61, 183
Smith, D. K., 26, 28–30, 168, 170, 179, 180
Smith, John, 118, 132
Sobel, Dana, 141, 180
Solin, 33
Soriano, Osvaldo, 121, 143, 178
Stace, 27, 54
Steinthal, Heymann, 112
Stevenson, Robert Louis, 155
Strabo, 60, 99, 116
Straet, Jan ver der, 120
Suleyman, Mansa, 85

Theodosius (Emperor), 14
Theophilus of Ferrariis, 123
Thevet, André, 144
Tiana, Rodrigo de, 79
Todorov, Tzvetan, 103
Tolkien, J. R. R., 155
Tolstoy, Leo, 56
Totti, Francesco, 14
Trevisa, John, 57, 61
Tudèle, Benjamin de, 32, 36, 168
Turchi, Peter, 148, 181
Tyr, Maxime de, 38

Umarī, Chihab al-, 83, 85–86, 174
Uys, Jamie, 9

Van Bommel, Mark, 122
Van Herk, Aritha, 55, 155–56, 171, 182

Van Sertima, Ivan, 84, 89
Varenius, Bernhardus, 58
Varron, 12–13, 166
Vattimo, Gianni, 165
Verrazzano, Giovanni da, 130
Verthema, Ludovico de, 36
Vespucci, Amerigo, 118–20, 127–28, 130–31
Veyne, Paul, 13
Victor, Gaius Julius, 110
Virgil, 27, 51, 53, 68, 75, 100, 171
Vivaldi, Ugolino, 56–57, 86
Vivaldi, Vadino, 56–57, 86
Volchok, Arkady, 135
Voltaire (François-Marie Arouet), 143

Wace, Robert, 39
Waldseemüller, Martin, 127–28
Walker, William, 58–59, 121, 172
Weil, Simon, 18
Wiener, Leo, 175
Wilcock, Juan Rodolfo, 17
Wordsworth, William, 133

Ximénez, Fortún, 117

Yáñez, Vicente, 79
Yerasimos, Stéphane, 81, 174
Yvernault, Martine, xvii

Zeno, Antonio, 61–62
Zeno, Carlo, 62
Zeno, Nicolò, 61–62
Zhanji (Xuande), Zhu, 91
Zhu Di (Yongle Emperor), 90
Zichmni, 61–62
Žižek, Slavoj, 161, 183
Zulini, Valerio, 46
Zumthor, Paul, 25, 38–39, 44–46, 99–100, 107, 144, 168, 170, 176–77, 180
Zweig, Stefan, 127–28, 179

Milton Keynes UK
Ingram Content Group UK Ltd.
UKHW012253010224
437121UK00006B/262